Studies in Computational Intelligence

Volume 676

Series editor

Janusz Kacprzyk, Polish Academy of Sciences, Warsaw, Poland
e-mail: kacprzyk@ibspan.waw.pl

About this Series

The series "Studies in Computational Intelligence" (SCI) publishes new developments and advances in the various areas of computational intelligence—quickly and with a high quality. The intent is to cover the theory, applications, and design methods of computational intelligence, as embedded in the fields of engineering, computer science, physics and life sciences, as well as the methodologies behind them. The series contains monographs, lecture notes and edited volumes in computational intelligence spanning the areas of neural networks, connectionist systems, genetic algorithms, evolutionary computation, artificial intelligence, cellular automata, self-organizing systems, soft computing, fuzzy systems, and hybrid intelligent systems. Of particular value to both the contributors and the readership are the short publication timeframe and the worldwide distribution, which enable both wide and rapid dissemination of research output.

More information about this series at http://www.springer.com/series/7092

Ajith Abraham · Rafael Falcon
Mario Koeppen
Editors

Computational Intelligence in Wireless Sensor Networks

Recent Advances and Future Challenges

 Springer

Editors
Ajith Abraham
Scientific Network for Innovation
  and Research Excellence
Machine Intelligence Research Labs
  (MIR Labs)
Auburn, WA
USA

Mario Koeppen
Department of Computer Science
  and Electronics
Graduate School of Creative Informatics
Kyushu Institute of Technology
Fukuoka
Japan

Rafael Falcon
Research and Engineering Division
Larus Technologies
Ottawa, ON
Canada

and

School of Electrical Engineering
  and Computer Science
University of Ottawa
Ottawa, ON
Canada

ISSN 1860-949X   ISSN 1860-9503   (electronic)
Studies in Computational Intelligence
ISBN 978-3-319-83804-5   ISBN 978-3-319-47715-2   (eBook)
DOI 10.1007/978-3-319-47715-2

Printed on acid-free paper

This Springer imprint is published by Springer Nature
The registered company is Springer International Publishing AG
The registered company address is: Gewerbestrasse 11, 6330 Cham, Switzerland

Preface

Wireless Sensor Networks (WSNs) are rapidly becoming a technological corner-stone for modern societies. These collections of autonomous and distributed nodes capable of sensing, communication, processing, and even self-organization continue to earn notoriety as they serve as the backbone of emerging intelligent information-driven paradigms such as the *Internet of Things* [7, 12, 22], *Vehicular Clouds* [6, 19], or *Cyber-Physical Systems* [2, 4]. Over the last two decades, we have witnessed a plethora of developments related to theoretical innovations in WSNs that touch all aspects of their multilayered design, from more robust physical and medium access layers [23] to more efficient energy conservation [15, 18, 21] and self-organization protocols [5, 25]. The number of published surveys reporting successful WSN applications to dissimilar domains [1, 8–10, 20] is frankly overwhelming.

Computational Intelligence (CI) is a very active research discipline that encompasses a plethora of methodologies that draw inspiration from natural and social processes to model and solve a variety of challenging real-world problems [11, 13]. The appeal behind CI techniques revolves around the fact that they take into account the imprecise, vague, and uncertain knowledge that is often present in any realistic world model. Through the abstraction and simulation of intelligent systems such as bird flocks, fish schools, ant colonies, immune system cells, neural connections, and other highly parallel and distributed processes, the overhead imposed by the computational intractability of NP-hard optimization problems and, more recently, the emergence of Big Data [16], has been reasonably alleviated. The term CI is not indicative of a single methodology; rather, it describes a large umbrella under which several biologically and socially motivated techniques have emerged [11]. The CI field has outgrown its traditional foundations (centered around *artificial neural networks*, *fuzzy systems* and *evolutionary computation*) to embrace other related approaches that also pursue the same goals of tractability, robustness, and low solution cost [11, 13], including but not limited to: *rough sets*, *multi-valued logic*, *connectionist systems*, *swarm intelligence*, *artificial immune systems*, *granular computing*, *game theory*, *deep learning*, and the *hybridization* of the aforementioned systems.

CI techniques have much to offer to WSN in terms of the realization of periodical yet vital tasks such as *sensor node localization, data collection and aggregation, energy-aware routing/broadcasting*, and *sensor relocation* [14]. The interplay between both fields of study is growing in vitality and spills over other closely related areas such as *bio-inspired computing, robotics and vehicular systems*, thus crystallizing the foundations of an exciting multidisciplinary arena. *Bio-inspired networking* [3, 24] is a recently coined term that attempts to capture the impact of a large subset of CI methodologies to interconnected systems.

This volume is another initiative undertaken to emphasize the increasingly important role that CI methods are playing in solving a myriad of entangled WSN-related problems. The book serves as a guide for surveying several state-of-the art WSN scenarios in which CI approaches have been employed. The chapters in this volume do not offer an exhaustive picture of the rich landscape of CI-WSN applications given the breadth and depth of this interplay, with many problems rapidly arising as the pace of technology accelerates. The reader will find in this book how CI has contributed to solve a wide range of challenging problems, ranging from balancing the cost and accuracy of heterogeneous sensor deployments to recovering from real-time sensor failures to detecting attacks launched by malicious sensor nodes and enacting CI-based security schemes. Network managers, industry experts, academicians and practitioners alike (mostly in computer engineering, computer science, or applied mathematics) will benefit from the spectrum of successful applications reported in this volume. Senior undergraduate or graduate students may discover in this volume some problems well suited for their own research endeavors.

Volume Organization

Chapter 1 entitled "A Genetic Programming Approach to Cost-Sensitive Control in Wireless Sensor Networks" employs Genetic Programming (GP) to find suitable sensor control strategies that balance the accuracy of the measurements needed to monitor a certain region and the cost of powering these devices. In networks supporting multiple sensor types (a.k.a. heterogeneous WSNs), it is therefore desirable to develop cost-sensitive control algorithms that sample more expensive sensors only when necessary. The proposed solution has a twofold nature. First, a hierarchical method is proposed where GP solutions are sorted in a hierarchy of layers based on the cost of the sensors they use. Switching to the next more expensive layer takes place only if the prediction variance indicates uncertainty at lower layers. Second, the authors introduce non-hierarchical models that automatically select sensors based on both cost and accuracy. In experiments using a synthesized dataset and ten real datasets, the hierarchical method is shown to have significantly lower prediction costs than the non-hierarchical method.

Wireless Mesh Networks (WMNs) are a particular type of WSN whose topology can vary from a simple star network to an advanced multi-hop one. The main topological feature is that nodes are organized in a mesh topology, thus making WMNs a reliable infrastructure through the redundancy of multi-hop communications. In Chapter 2 "A Study on Performance of Hill Climbing Heuristic Method for Router Placement in Wireless Mesh Networks", the authors put forth an approach based on Hill Climbing (HC), a simple local search method, to quickly identify near-optimal router locations in a WMN so as to improve its Quality of Service (QoS) in terms of maximizing the network connectivity and client coverage. The ensuing bi-objective optimization problem is tackled via the HC heuristic method, whose performance is investigated under different distributions of client mesh nodes.

Chapter 3 titled "An Automated Irrigation System Based on a Low-Cost Microcontroller for Tomato Production in South India" introduces a practical result on a fuzzy logic-based irrigation controller for growing vegetables. The system consists of a feedback fuzzy logic controller that records key parameters with sensors, ZigbeeGPRS remote monitor, and a database. Based on the crop yield, the fuzzy logic controller acquires data from the sensors and applies fuzzy rules to determine a suitable irrigation time. A MaxMin inference engine and a Mamdani-type fuzzy inference system were adopted in order to make the best decision for each situation. The proposed system was developed and tested for the growth of tomato plants. It saves 50–60 % of the water utilization as well as the energy generation cost.

Chapter 4 "Artificial Neural Network Based Real-Time Urban Road Traffic State Estimation Framework" unveils a methodology that utilizes the existing cellular network infrastructure for road traffic data collection with a three-layer neural network model to estimate the complete link traffic state. The inputs to the neural network (NN) model include the probe vehicle's position, timestamps, and speeds. The framework integrates different modules that resort to different models in the process of traffic state estimation. Real A-GPS data gathered using A-GPS mobile phone on a moving vehicle on the set of chosen roads is used to evaluate the NN model. The trained NN is also used to estimate the road link speeds and compares them with ground truth speed (aggregate edge states) on a 10-min interval per hour. The estimation accuracy indicated that reliable link speed estimation can be generated and used to determine real-time urban road traffic conditions.

WSNs are subject to an ample range of potential attacks originated by malicious sensors. These attacks range from passive eavesdropping to active interfering and tampering of the communication. Chapter 5 "Attack Detection Using Evolutionary Computation" is concerned with the detection of such active attacks using the restricted capabilities of the sensor nodes. The underlying idea is that each sensor node is equipped with a simple intrusion detection system (IDS), hence an entire area can be monitored for malicious behavior in a distributed fashion. The automatic configuration of the IDS parameters is entrusted to *Multi-Objective Evolutionary Algorithms* (MOEAs) and illustrated via the selective forwarding attack and the delay attack. The proposed optimization framework provides Pareto front approximations consisting of different IDS settings with respect to three objectives, i.e., false positives, false negatives, and memory consumption. Furthermore, the authors discuss various attacker strategies and the robustness of the IDS settings found for a specific attacker strategy in cases where another attacker strategy is enacted.

Chapter 6 "Computational Intelligence Based Security in Wireless Sensor Networks: Technologies and Design Challenges" reviews the application of CI techniques to developing security schemes for WSNs. Fuzzy sets, rough sets, neurocomputing and evolutionary approaches are among the formalisms that have been proposed to enable WSNs with security features. There is broad uncharted territory when it comes to designing CI-based security systems for WSNs.

Wireless Visual Sensor Networks (WVSNs) are a type of WSNs that are heavily used for sensitive applications such as video surveillance and monitoring. To overcome the typical constraints of a WVSN in terms of its limited memory, energy, and bandwidth, Compressed Sensing (CS) techniques are brought into place with the aim of reconstructing sparse signals using very few measurements. Anomaly detection can then be accomplished in a more efficient manner using CS. Chapter 7 "Efficient Anomaly Detection System for Video Surveillance Application in WVSN with Particle Swarm Optimization" employs the popular Particle Swarm Optimization (PSO) metaheuristic algorithm to optimize the minimum number of compressed measurements and the routing of the information towards the destination. The proposed system is capable of detecting targets with fewer measurements and transmitting the required compressive measurements for reconstruction with less energy, thereby increasing the network lifetime.

Mobile robots are brought into a WSN to perform a wide range of tasks that optimize the WSN operation and extend its lifetime. One example of this is the replacement of damaged sensors with other functional, passive ones already deployed in the monitoring region. This problem has been recently studied under the name of *Robot-Assisted Sensor Relocation* (RASR) and cast as a combinatorial optimization problem. Chapter 8 entitled "Planning Robust Sensor Relocation Trajectories for a Mobile Robot with Evolutionary Multi-objective Optimization" extends the previous RASR formulation by actively considering the current energy levels of the participating passive sensors as well as the ideal locations for their deployment as additional decision objectives. This results in more robust sensor relocation trajectories to be pursued by the mobile robot. The authors explore six prominent MOEA implementations and discuss their performance with WSNs of varying sizes, inflicted damage levels, and passive sensor densities. They also tailor a recently proposed Risk Management Framework to proactively detect sensors that are at a high risk for failure and replace them before any network coverage is lost.

Future Challenges

The new generation of wireless networking involving the Internet of Things, Cyber Physical Systems etc., will result in *higher rate integrated communications*. Understanding and managing the complexity of such networks' bandwidth, capacity, security and Quality of Service (QoS) requirements will all be significant research challenges.

Currently we are also experiencing an *explosion of mobile data traffic*, characterized by the 4 V's Big Data vector: volume, velocity, variety, and veracity [16]. So, designing suitable frameworks to handle such Big Data in a wireless environment using appropriate Computational Intelligence tools will be a real challenge. Important aspects revolve around *real-time distributed control, processing and visualization* of these data streams in order to generate *actionable intelligence* that can better assist the decision-making process. A *risk-aware view* [17] of the WSN-monitored environment is not only beneficial but necessary in order to emphasize on the events of interest and declutter the operator's workspace.

We hope that the suite of technical contributions gathered in this book help drive further momentum into many theoretical and practical aspects of the wonderful synergy between CI methods and the WSN realm. Enjoy the reading!

Auburn, USA   Ajith Abraham
Ottawa, Canada   Rafael Falcon
Kyushu, Japan   Mario Koeppen
May 2016

References

1. Abbasi, A.Z., Islam, N., Shaikh, Z.A., et al.: A review of wireless sensors and networks' applications in agriculture. Comput. Stand. Interfaces **36**(2), 263–270 (2014)
2. Alur, R.: Principles of Cyber-Physical Systems. MIT Press, Cambridge (2015)
3. Câmara, D.: Bio-Inspired Networking. Elsevier, Amsterdam (2015)
4. Derler, P., Lee, E.A., Vincentelli, A.S.: Modeling Cyber–Physical Systems. Proc. IEEE **100** (1), 13–28 (2012)
5. Falcon Martinez, R.J.: Towards fault reactiveness in wireless sensor networks with mobile carrier robots. Ph.D. thesis, Université d'Ottawa/University of Ottawa (2012)
6. Gerla, M., Lee, E.K., Pau, G., Lee, U.: Internet of vehicles: from intelligent grid to autonomous cars and vehicular clouds. In: 2014 IEEE World Forum on Internet of Things (WF-IoT), 2014 pp. 241–246 (2014)
7. Gubbi, J., Buyya, R., Marusic, S., Palaniswami, M.: Internet of Things (IoT): A vision, architectural elements, and future directions. Future Gener. Comput. Syst. **29**(7), 1645–1660 (2013)
8. Güngör, V.Ç., Hancke, G.P.: Industrial Wireless Sensor Networks: Applications, Protocols, and Standards. CRC Press (2013)
9. Hadjidj, A., Souil, M., Bouabdallah, A., Challal, Y., Owen, H.: Wireless sensor networks for rehabilitation applications: challenges and opportunities. J. Netw. Comput. Appl. **36**(1), 1–15 (2013)
10. Heidemann, J., Stojanovic, M., Zorzi, M.: Underwater sensor networks: applications, advances and challenges. Phil. Trans. R. Soc. A **370**(1958), 158–175 (2012)
11. Kacprzyk, J., Pedrycz, W.: Springer Handbook of Computational Intelligence. Springer, New York (2015)
12. Kopetz, H.: Internet of things. In: Real-time Systems, pp. 307–323. Springer, New York (2011)

13. Kruse, R., Borgelt, C., Klawonn, F., Moewes, C., Steinbrecher, M., Held, P.: Computational Intelligence: A Methodological Introduction. Springer Science & Business Media, London (2013)
14. Kulkarni, R.V., Forster, A., Venayagamoorthy, G.K.: Computational intelligence in wireless sensor networks: a survey. IEEE Commun. Surv. Tutor. **13**(1), 68–96 (2011)
15. Liu, X.: A survey on clustering routing protocols in wireless sensor networks. Sens. **12**(8), 11113–11153 (2012)
16. Mayer-Schönberger, V., Cukier, K.: Big data: A Revolution That Will Transform How We Live, Work, and Think. Houghton Mifflin Harcourt, Boston (2013)
17. McCausland, J., Abielmona, R., Falcon, R., Petriu, E.: Risk-aware sensor networks for critical infrastructure monitoring. In: 11th International Conference on Shock & Impact Loads on Structures (SILOS). Ottawa, Canada (2015)
18. Nikolidakis, S.A., Kandris, D., Vergados, D.D., Douligeris, C.: Energy efficient routing in wireless sensor networks through balanced clustering. Algorithms **6**(1), 29–42 (2013)
19. Olariu, S., Hristov, T., Yan, G.: The next paradigm shift: from vehicular networks to vehicular clouds. Second Edn., Mobile Ad Hoc Networking: Cutting Edge Directions, pp. 645–700 (2013)
20. Othman, M.F., Shazali, K.: Wireless sensor network applications: a study in environment monitoring system. Procedia Eng. **41**, 1204–1210 (2012)
21. Pantazis, N.A., Nikolidakis, S.A., Vergados, D.D.: Energy-efficient routing protocols in wireless sensor networks: a survey. IEEE Commun. Surv. Tutor. **15**(2), 551–591 (2013)
22. Perera, C., Zaslavsky, A., Christen, P., Georgakopoulos, D.: Context aware computing for the internet of things: a survey. IEEE Commun. Surv. Tutor. **16**(1), 414–454 (2014)
23. Suriyachai, P., Roedig, U., Scott, A.: A survey of MAC protocols for mission-critical applications in wireless sensor networks. IEEE Commun. Surv. Tutor. **14**(2), 240–264 (2012)
24. Xiao, Y.: Bio-inspired Computing and Networking. CRC Press, Taylor & Francis (2016)
25. Younis, M., Senturk, I.F., Akkaya, K., Lee, S., Senel, F.: Topology management techniques for tolerating node failures in wireless sensor networks: a survey. Comput. Netw. **58**, 254–283 (2014)

Contents

A Genetic Programming Approach to Cost-Sensitive Control in Wireless Sensor Networks

Afsoon Yousefi Zowj, Josh C. Bongard and Christian Skalka

Abstract In some wireless sensor network applications, multiple sensors can be used to measure the same variable, while differing in their sampling *cost*, for example in their power requirements. This raises the problem of automatically controlling heterogeneous sensor suites in wireless sensor network applications, in a manner that balances cost and accuracy of sensors. We apply genetic programming (GP) to this problem, considering two basic approaches. First, we construct a hierarchy of models, where increasing levels in the hierarchy use sensors of increasing cost. If a model that polls low cost sensors exhibits too much prediction uncertainty, the burden of prediction is automatically transferred to a higher level model using more expensive sensors. Second, we train models with cost as an optimization objective, called non-hierarchical models, that use conditionals to automatically select sensors based on both cost and accuracy. We compare these approaches in a setting where the available budget for sampling is considered to remain constant, and in a setting where the system is sensitive to a fluctuating budget, for example available battery power. We show that in both settings, for increasingly challenging datasets, hierarchical models makes predictions with equivalent accuracy yet lower cost than non-hierarchical models.

1 Introduction

Wireless Sensor Networks (WSNs) have revolutionized environmental monitoring by combining low cost with flexibility in sensor capabilities [31]. They have been used in diverse environmental monitoring applications and continue to be adapted in new fields. Because WSNs are often, even typically, deployed in remote locations, and

A.Y. Zowj (✉) · J.C. Bongard · C. Skalka
Department of Computer Science, University of Vermont, Burlington, VT, USA
e-mail: ayousef1@uvm.edu

J.C. Bongard
e-mail: jbongard@uvm.edu

C. Skalka
e-mail: ceskalka@uvm.edu

© Springer International Publishing AG 2017   1
A. Abraham et al. (eds.), *Computational Intelligence in Wireless Sensor Networks*,
Studies in Computational Intelligence SCI 676, DOI 10.1007/978-3-319-47715-2_1

thus rely on combinations of battery power and energy harvesting, a major challenge in WSN design is to minimize system power consumption.

Minimizing power consumption can be accomplished in a variety of ways, in particular by adapting sensor control strategies that optimize the balance between measurement accuracy and the cost of powering sensors [30]. In this paper, we propose new sensor control algorithms for WSNs with heterogeneous sensor suites that balance cost and accuracy, obtained using genetic programming (GP) techniques.

By "heterogeneous sensor suite", we mean WSNs equipped with multiple types of sensors for prediction of the same phenomena. Each of these sensors is characterized by its accuracy in relation to the phenomena, and a cost of use which is often measured by its power consumption. Such systems support multi-modal sensor fusion, a well-studied technique where data from multiple sensor modalities (types) is combined to predict a single variable [30]. The contribution of our work is a consideration of cost in multi-modal sensor fusion, and the development and testing of associated control algorithms. These algorithms will call upon particular sensors only when needed, and otherwise rely on the cheapest available sensors at any given time. Our problem is distinguished from adaptive sampling [30] in that the latter is concerned with optimally modulating sampling frequency of a given sensor, not choosing between a suite of possible sensors.

While various multi-modal sensor fusion applications exist, we are especially interested in the Snowcloud system which combines snow density telemetry with snow depth and air temperature sensors to predict areal snow water equivalent (SWE) [24]. We envision extending Snowcloud to incorporate ground based light detection and ranging (LIDAR) scanning [5] to be used for SWE estimation as part of its sensor suite. However, while LIDAR yields more accurate data than existing Snowcloud telemetry, it does so at significant additional power cost. Thus, the challenge is to commit these resources only at optimal times. It is also a refinement of multi-modal sensor fusion, since we are mainly interested in settings where available data gathering techniques differ in accuracy, with less accurate sensors being cheaper than more accurate ones.

A fundamental component of our approach is the use of prediction *uncertainty* to drive sensor usage. We propose a scheme whereby predictions are attempted using lower-cost sensors at first. If uncertainty is below an acceptable threshold, then the prediction is used. Otherwise we switch to higher-cost sensors, make a new prediction based on those inputs, evaluate uncertainty again, and continue to move the burden of prediction to more accurate and costly sensors as needed. This scheme is discussed in detail in Sect. 2.4 and described graphically in Fig. 2. Note that while the Snowcloud system is an intended application of this scheme, it can be generalized to any WSN application using heterogeneous sensor suites comprising sensors with varying cost and accuracy.

To quantify uncertainty we are aided by machine learning ensemble methods—we use entropy in ensemble predictions as a proxy for uncertainty [23]. To obtain predictive models themselves, in this work we use genetic programming (GP) [14]. This is largely due to characteristics of our intended application space. Previous work has demonstrated that the relationships between snow cover and the topographic and meteorological factors that influence it include non-linearities [26], while the spatial distribution of SWE is nonlinear because it is influenced simultaneously by various forcing effects [27]. Nonlinear predictors are therefore desirable. Furthermore, recent results [7] show that GP has advantages over other approaches (such as decision trees) due to associated techniques for preventing overfitting, e.g. treating model size minimization as an objective [12]. Although C4.5 only supports classification, sufficiently fine classification granularity can achieve competitive performance on regression problems, and this approach is popular in the environmental science community [7]. Finally, GP is appealing due to its white-box nature: it can potentially provide physical insights into modeled phenomena.

An alternative approach to our problem is to not rely on external measures of entropy to switch between sensors, but to treat cost as an additional objective in a multi-objective optimization problem. We explore this option in our work, in direct comparison to the hierarchical approach. However, due to the "curse of dimensionality", adding another optimization dimension may have deleterious effects on prediction performance, especially since selection for size to avoid overfitting already imposes a multi-objective optimization regime [6]. We therefore hypothesize that a hierarchical approach will outperform a non-hierarchical approach in settings with multiple sensors of differing predictive abilities, and we explore this comparison in our experiments.

In our initial comparison of these two approaches—hierarchical and non-hierarchical—our regime is not concerned with the available budget. However, in real deployments, budget levels can have significant impacts on what sensors are chosen. For example, if battery levels are low, expensive sensors should probably be avoided regardless of prediction uncertainty, both to reduce system downtime and sensor noise. Therefore, we also consider a comparison of the hierarchical and non-hierarchical approaches in a setting where models are sensitive to dynamic budget fluctuations. As for the basic setting, we hypothesize that the hierarchical approach will perform better than the non-hierarchical.

1.1 Related Work

Previous work on adaptive sampling [30] has aimed to reduce sampling rates in Resource Constrained Sensor Systems (RCSS) applications to balance sensor cost and accuracy. In particular, Alippi et al. [4] have tried to find the optimal adaptive frequency of sampling for avalanche monitoring. It has further been claimed that compressed sensing—sending aggregated data instead of raw data—performs better in conjunction with reducing sampling rates, rather than just reducing the sampling rate alone [17]. A variety of methods for compressed sensing [8] have been proposed. Although these methods have achieved cost reduction in monitoring, they are not applicable to our problem since we intend not to change the rate of sampling of one sensor type, but rather to reduce sampling cost by switching between available sensors of different type and accuracy.

Another line of work focuses on finding the optimal location for sensors in distributed deployments, in order to maximize accuracy while minimizing deployment densities. Krause et al. [15] have used a probabilistic method to predict the communication cost for a given deployment topology. Papadimitriou et al. [19] have employed GP and a Bayesian statistical method to minimize entropy over a set of sensor locations. In contrast, our work is concerned with reducing the cost of sampling from an available set of sensors at any given time, not with reducing the densities of sensor topologies.

In work on so-called multi-modal sensor fusion, data from multiple sensors in a potentially heterogeneous suite are aggregated to monitor a specific measurement application [9, 28]. This method has been widely used, for example in visual monitoring [18, 20] and target tracking [21, 25]. Data fusion focuses on sensor applications that need to compute the correlation between multiple sensor modules and cannot be measured by a single sensor. However, these works do not consider the cost of using different sensors, or minimizing cost.

Cost sensitive multi-modal sensor fusion methods have been developed to balance cost against accuracy, with an eye towards providing fault tolerance [13]. However, we are not concerned with fault tolerance, but strictly between selecting sensors from heterogeneous suites. Willett et al. [30] use a small number of sensors to send their readings to a fusion center, and based on the correlation among the sensed data, the fusion center decides which additional sensors should be activated. The same concept has also been tried in a distributed fashion [16]. However, sensing costs in these cases are a function of the number of sensors sampled, not their type.

Perhaps most related to our work is that of Wang et al. [29]. They propose a method to find the optimal set of sensors to be polled, using a hybrid tree, where non-leaf nodes act as a decision tree and leaves are standard regression models using a subset of sensors. However, these trees support decision making based on external constraints, i.e., which sensors to use depending on an organization's goals and resources. In contrast, our models are intended to support automated sensor control in WSNs during deployments.

Outside of the adaptive sampling and sensor fusion fields, multi-objective optimization has been used for cost-sensitive modeling. For example Kim [12] sets error as one objective and tree size as another, as we do here. Zhao [32] sets the false negative rate and false positive rate as the two objectives. However, these works do not consider the hierarchical approach that we do.

1.2   Organization of the Chapter

The remaining text is organized as follows. In Sect. 2 we formalize our basic problem description, and explain how hierarchical and non-hierarchical models are constructed. In Sect. 3 we describe the experiments we perform to compare these two approaches, and the quantitative results from those experiments. In Sect. 4 we describe an extension where dynamically changing budget information can be taken into account, and reformulate a problem formalization, as well as a description of methods, experiments, and quantitative results in this extended setting. In Sect. 5 we discuss and reflect on our quantitative results for all experiments. In Sect. 6 we conclude with remarks on future work.

2 Methods

This section provides a formalization of the problem, how genetic programming is applied to solve it, and the two variants of genetic programming that we compare in this work. All of the material for replicating the work described here is available online [1].

2.1 Problem Formalization

Let us assume that t values of some environmental phenomenon $\mathbf{g}$ (the ground truth) are known at time steps $1, \ldots t$. These values are stored in $\mathbf{g} = g_1, \ldots g_t$. Let us further assume there are k sensors $s_1, \ldots s_k$ available that can be used to predict $\mathbf{g}$. Let $r_i^{(t)}$ denote the reading of sensor i taken at time t. Moreover, let $s^{(t)}$ and $r^{(t)}$ denote a subset of sensors, and readings taken from them, at time t. We denote the amount of variance of $\mathbf{g}$ explained by sensor i as $v_{r_i}^{(\mathbf{g})}$. This value is determined by linearly regressing only r_i against g. Finally, let $e_i = 100(1 - v_{r_i}^{(\mathbf{g})})$ and c_i represent the prediction error and cost of using sensor i respectively. Using this formulation, e_i represents the percentage of prediction error incurred by just using sensor i to predict $\mathbf{g}$.

The cost of a sensor c_i is usually inversely proportional to its error e_i, so for the work reported below, we set $c_i = v_{r_i}^{(\mathbf{g})}$ for each sensor. In certain sensor deployments there may be other factors that affect c_i such as power consumption, market price, effort required to collect a sensor's reading, proprietary issues, and so on. In any case it is important to clarify that in this work we only consider costs of sensor sampling, not operational costs of the platform, e.g. the cost of post-sampling data processing.

We suppose that an ordering of sensors exists such that s_1 is the least expensive sensor with the highest error and s_k is the most expensive sensor with the lowest error. Formally,

$$\forall i, j . 1 \leq i < j \leq k \rightarrow e_i > e_j \wedge c_i < c_j.$$

Let us denote the prediction of a model using a subset of sensors at time t by $p^{(t)}$, i.e., $p^{(t)}$ is a function on $\mathbf{r}^{(t)}$. Then, the error of each sampling $e^{(t)}$ would be

$$e^{(t)} \triangleq |p^{(t)} - g^{(t)}|.$$

The cost of each sampling, $c^{(t)}$ is the cumulated cost of all sensors $s_i \in \mathbf{s}^{(t)}$ that were polled for that sampling:

$$c^{(t)} \triangleq \sum_{j \in \{i \mid s_i \in \mathbf{s}^{(t)}\}} c_j.$$

It is desired that each sampling $\mathbf{s}^{(t)}$ entails low error and cost. That is, the following equality is desirable:

$$\operatorname*{argmin}_{\mathbf{s}^{(t)}} e^{(t)} = \operatorname*{argmin}_{\mathbf{s}^{(t)}} c^{(t)}.$$

Our goal is to design models which combine and transform sensor readings to accurately predict the outcome measure, but can also intelligently determine which sensors to poll when cheap, less accurate sensors exhibit uncertainty about the current prediction.

2.2 General Genetic Programming Approach

Genetic programming has widely been employed for regression tasks in which the functional form of the equations relating inputs to outputs is unknown [14]. Here, inputs are sensor values and the output is a prediction for a given outcome measurement.

Although many recent improvements have been proposed for GP, here we have kept the genetic programming algorithm simple and instead focused on comparing GP-generated hierarchical and non-hierarchical models. Thus, GP is restricted to the four simple algebraic operators, and each evolutionary trial is initialized with a fixed-sized population of 100 randomly-generated solutions containing three nodes. Maximum tree depth is not set since the tree size is considered as an objective in multi-objective optimization. The crossover rate is set to 0.2 and no fitness stall is considered. If the number of non-dominated solutions reaches 50 % of the population size, the training restarts. At the conclusion of each generation, four values are computed for each solution: (1) *error* on training data as defined below, (2) the combined *cost* of the sensors used to make the prediction, (3) the *size* of the solution, and (4) the *age* of the solution. We now discuss each in turn.

Error Let n be the population size and j range over $\{1, \ldots, n\}$. Let t_j be some solution tree. We represent the error of sampling at time t using solution t_j with $e_{t_j}^{(t)}$. Moreover, $d^{(\text{train})}$ and $d^{(\text{test})}$ denote the training dataset and testing dataset, respectively. Then, we define the error on training data using solution t_j by $e_{t_j}^{(\text{train})}$ and as the average of $e_{t_j}^{(t)}$ on all samples in $d^{(\text{train})}$, i.e.,

$$e_{t_j}^{(\text{train})} \triangleq \sum_{g^{(t)} \in d^{(\text{train})}} \frac{e_{t_j}^{(t)}}{|d^{(\text{train})}|}. \tag{1}$$

Each solution t_j was allowed to use a subset (possibly empty) of available sensors. The cost of each solution depends on the sensors that are employed and the sampling.

Cost As described in the following sub-sections, current sensor readings may trigger readings from additional sensors. Thus, different $r_i^{(t)}$ may cause t_j to need different $\mathbf{s}^{(t)}$. The average cost of a tree on training data $c_{t_j}^{(\text{train})}$ is thus defined as the cost of all of the sensors that have been used to predict the outcome for each training instance, averaged over all instances in the training dataset:

$$
c_{t_j}^{(\text{train})} \triangleq \sum_{\mathbf{r}^{(t)} \in d^{(\text{train})}} \sum_{l \in \{i \mid s_i \in \mathbf{s}^{(t)}\}} \frac{c_l}{|d^{(\text{train})}|}. \tag{2}
$$

If a solution uses a sensor more than once, no extra cost is incurred: because the sensor has already been polled, its output is already available and can thus be re-used as often as required.

Size To avoid bloat, solution size, defined as the number of nodes in the tree, was incorporated into the fitness objectives during the optimization process [11].

Age We employed the Age-Fitness Pareto Optimization (AFPO) method [22], which injects a new randomly-generated solution into the population at each generation and compares the solutions with same age in an effort to guard against convergence. Each solution's age is defined as the number of generations since its oldest ancestor was injected into the population. A new solution produced by mutating an existing solution inherits the same age as its parent. If two existing parents are crossed to produce two new offspring, the offspring inherit the age of the older of the two parents. AFPO is a multiobjective optimization method as solution age is used as an additional fitness objective during optimization.

Optimization At the end of each generation, the Pareto front is computed according to the objectives used, and the dominated solutions are discarded. Multi-objective optimization with all four objectives described above could easily lead to population collapse in the sense that all members of the population could become non-dominated. To guard against this eventuality, one possibility is to restart the evolutionary run with new solutions if no dominated solutions are detected in the population at the end of a given generation. Alternatively, a very large population size can be employed. However, both of these solutions greatly increase the computational effort required to obtain satisfactory solutions to the given problem. To avoid this situation, different multi-objective optimization approaches has been proposed. One of the simplest non-parametric approaches is to reduce the number of objectives by multiplying objectives together and using the result in the optimization process [10]. In this experiment, since error is the most important outcome, error is used for the primary objective and the second objective is the result of multiplying cost, size and age together.

Once the dominated solutions are deleted, the empty slots in the population are then filled by mutating and crossing copies of the non-dominated solutions. Tournament selection is used to select parents from the front for these operations. After the last generation, age is discarded when computing members of the Pareto front, since the goal is to use only small, accurate and cost-effective solutions for prediction, regardless of their age.

Fig. 1 a Non-hierarchical framework. **b** A non-hierarchical sample solution

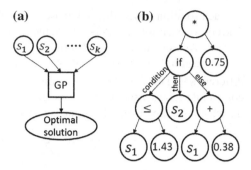

2.3   Non-hierarchical GP

A naive approach to cost-sensitive modeling using GP would be to evolve individual trees that add conditional and comparative operators to the base set of operators, and allow the tree to poll the values of all sensors if desired, as shown in Fig. 1a. In this way, different parts of the solution tree will be visited depending on the current values of the sensors. If less expensive sensors report a certain combination of values which in the current circumstances is unlikely to provide a good prediction, successful solutions may evolve that visit nodes containing references to expensive sensors.

Figure 1b shows an hypothetical example of a GP solution t_j that has evolved to encode a useful conditional. In this example, an inexpensive sensor s_1 is first polled. If its reported value $r_1^{(t)}$ is below some threshold, the reading of a more expensive sensor s_2 will be used. It is assumed here that s_1 leads to making poor predictions of the outcome if its reading is below 1.43. If this threshold is exceeded, $r_1^{(t)}$ is then used to predict the outcome.

Conditional operators should, indirectly, encode the differential effects on the available sensors, and the relative costs of those sensors. Note that this is possible even if GP does not have direct access to these differential effects and costs, as they are indirectly reflected in the errors and costs incurred when each solution is evaluated. This issue is worth mentioning in that these effects are complex, non-linear and noisy, and even field experts cannot define them precisely.

2.4   Hierarchical GP

An alternative approach to reconciling prediction error and prediction cost is to build a hierarchy of models: models in the lower layers only have access to inexpensive sensors, while models in the upper layers have access to a greater subset of the sensors, including more expensive ones. When deployed, the overall model returns a prediction from a lower layer if the inexpensive sensors are confident of their combined prediction. If they are not, predictions are drawn from a higher layer.

Fig. 2 Hierarchical
framework (**a**). Using the
difference between training
data prediction variance and
test data prediction variance
as the condition for
switching between model
layers (**b**)

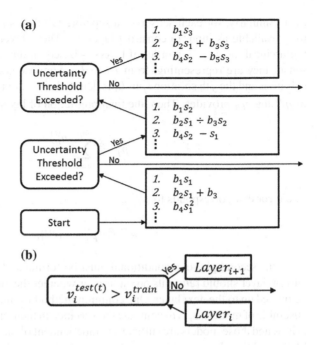

Briefly, constructing such a model proceeds in two phases:

1. Build a set of k layers, one for each sensor modality. For each layer i, run GP to find a set of accurate and low-cost solutions that use one or more sensors from the set $s_1, s_2, \ldots s_i$.
2. Define conditions which determine which layer should be allowed to provide the prediction, given the current environmental conditions.

Figure 2 illustrates what such a hierarchical model looks like. At the outset of attempting to provide a prediction for the current environmental conditions, the models stored in the lowest layer are evaluated, which only have access to the least expensive sensor s_1. If the certainty of their combined predictions is acceptable, return the combined prediction of these models. Otherwise, evaluate the models at the next layer, which have access to s_1 and the next least expensive sensor s_2. If these models are acceptably confident in the prediction, return their combined prediction; otherwise, evaluate the solutions at the next layer, and so on. If the top layer is reached, the combined predictions of the models found there are returned as the overall prediction, regardless of their level of certainty. The incremental construction of these models is described next.

Starting with the least expensive sensor s_1, GP is used to find the best models for converting $r_1^{(t)}$ to $g^{(t)}$. When GP terminates, the final non-dominated solutions are then organized as a group named layer L_1. The same process is repeated for s_2, except for the fact that since s_1 is already polled in L_1, it may be incorporated into models during evolution without incurring an extra cost for the solution tree that makes use of it. Similarly, for each sensor s_i, a separate GP run is performed with sensors s_1 to s_i available as input to construct layer L_i. These layers are then organized in a hierarchical fashion. The order of layers is based on the cost of the most expensive sensor they are representing, from L_1 to L_k. Suppose each layer L_i consists of n_i solutions and the jth solution t_j in L_i is denoted as $t_{i,j}$. Let $p_{t_{i,j}}^{(t)}$ denote the prediction of $g^{(t)}$ that $t_{i,j}$ provides. Then, the final prediction of layer L_i for $g^{(t)}$ is

$$p_{L_i}^{(t)} \triangleq \sum_{j=1}^{n_i} \frac{p_{t_{i,j}}^{(t)}}{n_i}.$$

The error that corresponds to $p_{L_i}^{(t)}$ is

$$e_{L_i}^{(t)} \triangleq |p_{L_i}^{(t)} - g^{(t)}|.$$

In the second phase, a conditional must be formulated to determine whether the current layer should return its prediction, or whether the burden of prediction should be passed up to the next layer. One common method for measuring how confident an ensemble of models is, is to compute the variance in their predictions [23]: if variance is low, and those models are sufficiently independent of one another, there is a greater likelihood that their combined predictions can be trusted. If variance is high, this is likely the result of differing assumptions encoded in the models, which cannot all be true reflections of the hidden relationship being modeled. Note the assumption here that the models are relatively independent: a set of identical models will never exhibit a variance in their predictions, regardless of how accurate the individual models are. We can be somewhat confident of the independence of our models, as they are produced by the AFPO algorithm: models with differing ages are likely to arrive on the final Pareto front used to build each layer, and such differently-aged genomes are likely to be independent because of their different genetic origins.

Formally: Let $p_{L_i}^{\text{train}(t)}$ and $e_{L_i}^{\text{train}(t)}$ denote $p_{L_i}^{(t)}$ and $e_{L_i}^{(t)}$ using $\mathbf{r}^{(t)}$ on d^{train}, respectively. Similarly, $p_{L_i}^{\text{test}(t)}$ and $e_{L_i}^{\text{test}(t)}$ respectively denote $p_{L_i}^{(t)}$ and $e_{L_i}^{(t)}$ using $\mathbf{r}^{(t)}$ on d^{test}. Moreover, assume $v_i^{\text{train}(t)}$ and $v_i^{\text{test}(t)}$ are the variances of all $p_{t_{i,j}}^{(t)}$s on d^{train} and d^{test}. Also, v_i^{train} denotes $v_i^{\text{train}(t)}$ averaged over all the samplings in d^{train}.

To determine whether the burden of prediction should remain with the current layer or passed off to a higher layer, we measure the difference in prediction variance between the models when presented with the training data (v_i^{train}) or with the testing data, i.e. the current environmental conditions ($v_i^{\text{test}(t)}$). When $v_i^{\text{test}(t)}$ is almost the same as v_i^{train}, there is a high probability that $e_{L_i}^{\text{test}(t)}$ is an approximation of $e_{L_i}^{\text{train}(t)}$, and we can be relatively confident that these models will yield a good collective prediction of $g^{(t)}$. When the variance of test data prediction is significantly higher than prediction on the training data, this signals that the solutions in that layer are exhibiting increased disagreement regarding the current environmental conditions.

This could be due to the fact that a specific sensor is not physically able to predict under the current conditions, or the solutions have not been trained for the current situation. In such an eventuality it would be advantageous to switch to the next layer, in the hope that its models will exhibit more confidence in their ability to predict the current conditions. In this paper, the variance is considered as a proxy for entropy, but any other entropy related metric could be used instead. Figure 2 illustrates how this intuition is encoded into the switching condition in the hierarchy of layers.

By considering the amount of difference between prediction variance on training and testing data, we can dynamically tune how conservative or liberal the overall hierarchical model is: if little difference is tolerated, the burden of prediction will often be passed to higher layers, resulting in expensive yet accurate predictions; if much difference is tolerated, lower levels will tend to predict, resulting in less expensive and less accurate predictions. The advantage of this approach is that the amount of tolerance could be dynamically tuned based on the current available budget for sensing.

For example, for larger budgets, more cost could be expended in order to obtain more accurate results. In this regard, the tolerance of the difference between variances could be decreased, transferring the burden of prediction to higher layers. Similarly for small budgets, the tolerance would be increased. Through this adjustment, more disagreement would be tolerated and less accurate predictions would be obtained for lower cost. To implement this dynamic tuning given a fluctuating budget, a tolerance parameter $\tau \in [0, 1]$ is defined, reflecting the tolerance of disagreement between the solutions of a given layer. Equation (3) demonstrates how this parameter is used to determine which level should be activated for prediction.

$$p^{(t)} = \begin{cases} p_{L_i}^{(t)} & \text{if } v_i^{\text{train}} > |1 - \tau| \cdot v_i^{\text{test}(t)} \\ p_{L_{i+1}}^{(t)} & \text{otherwise} \end{cases} \tag{3}$$

It should be noted that in the present work, the same value for τ is used at the interstices between each pair of layers. However, different values for τ could be employed between different layers to enable the model to respond better to changes in the overall available budget. The extreme cases occur when $\tau = 0$ or $\tau = 1$. The former case ensures that the condition in Eq. 3 holds when the prediction variance on the testing data is greater than the prediction variance on the training data. This occurs with high probability, so setting $\tau = 0$ tends to extract the predictions from solutions on the uppermost layer. Setting $\tau = 1$ ensures that the first layer always provides the prediction since variance on the testing data will always be finite. Values greater than $\tau = 1$ are not investigated in this work, but are possible. Greater τ value increases the probability of the condition to be true. $\tau = \infty$ causes the condition to always be true, thus the method always collects predictions from the last layer.

3 Results

The proposed methods are evaluated over two set of experiments, using a synthesized dataset and ten actual datasets. This section summarizes these datasets, experimental setups, and quantitative results. These results were also reported in a preliminary version of the work reported here [33].

3.1 Synthesized Data

In these experiment, the proposed methods have been evaluated on a synthetic system monitored by three different sensors. Table 1 shows these three sensors, their readings in relation to $g^{(t)}$, and their cost.

To create the training and testing datasets, at first coefficients in the equations of the sensor relations, i.e., $b_{i,j}$, were randomly selected in the range [0, 1]. Then, random numbers were generated for $g^{(t)}$ in the range [0, 3], and used to calculate the sensor readings based on the given template and selected coefficients. The training and testing dataset sizes were 150 and 50, respectively, and each experiment was repeated 40 times.

Non-hierarchical setup The population size is 100 and is trained for 300 generations. The optimization process during the last generation does not consider *age* as an objective and the Pareto front is selected using *error* and *cost × size* as two separate objectives. After training, the knee of the non-dominated solutions is selected and tested using the testing dataset. In order to select the knee, the euclidean distance of each solution on the Pareto front is calculated from the ideal model. The ideal model is a solution with no error and zero cost. This is defined as follows:

$$t_{\text{knee}} = \underset{t_j \in \text{Pareto front}}{\text{argmin}} \sqrt{(e_{t_j} - 0)^2 + (c_{t_j} - 0)^2}.$$

Hierarchical setup The population size for each layer is 100 and each layer was trained for 100 generations to equalize the total computational effort applied in both methods. Similarly to the non-hierarchical setup, during the last generation, *age* is not considered in the Pareto optimization process, and non-dominated solutions are selected based on *error* and *cost × size* as two separate objectives. After training, for

Table 1 Available sensors and their features

Name	Equation template of $r_i^{(t)}$	Cost
s_3	$g^{(t)}$	0.3
s_2	$b_{2,1}g^{(t)} + b_{2,2}$	0.2
s_1	$b_{1,1}(g^{(t)})^2 + b_{1,2}g^{(t)} + b_{1,3}$	0.1

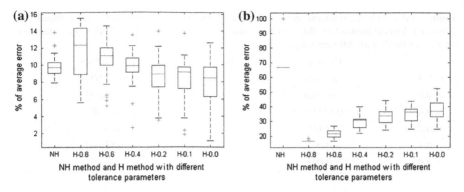

Fig. 3 Average error (**a**) and average cost (**b**) on the test data for the non-hierarchical and the hierarchical methods with different tolerance parameters. Statistical significance of these results are reported in Table 2

each layer L_i, the variance of the solutions output on training data v_i^{train} is computed and stored as the threshold of switching to the next layer L_{i+1}. This variance is not computed for the layer corresponding to the most expensive sensor, i.e., L_3, since there are no more sensors to be called. The experiment was repeated 40 times for each of the different tolerance parameters $\tau = 0.0, 0.1, 0.2, 0.4, 0.6, 0.8$.

Results on Synthesized Data We now consider average error and cost of the different modeling approaches on synthesized data and report P-values for two tailed t-tests where $\alpha = 0.5$.

Average error The average error of the non-hierarchical method is $e_{t_j}^{\text{test}}$, where t_j is the final selected solution. The average error of the hierarchical method is the average of $e_{L_i}^{\text{test}}$, where L_i is the last layer reached in the hierarchy, during the sampling. As can be seen in Fig. 3, the largest difference in error occurs at maximum tolerance i.e. $\tau = 0.8$ where the error of the hierarchical method is 1.34 % higher than the non-hierarchical method. The hierarchical method tends to achieve lower average error when the tolerance parameter is $\tau < 0.4$. P-values obtained for different tolerance parameters are represented in Table 2a and show that $\tau = 0.4$ is the boundary where the hierarchical method begins to outperform the non-hierarchical method.

Average cost By considering t_j as the final selected solution in the non-hierarchical method, the average cost is $c_{t_j}^{\text{test}}$. The average cost of the hierarchical method is the average of $\sum_{j=1}^{i} c_{L_j}^{\text{test}}$, where the last layer reached during the sampling is L_i. In order to compare both methods and understand how much of the potential cost each method uses, the cost of each method is represented as the percentage of cost of using all available sensors. Figure 3 shows that the average cost of the hierarchical method is significantly lower than the non-hierarchical method (at most 54.88 % and at least 33.81 % lower cost). Table 2b summarizes the p-values to show how significantly the cost of the hierarchical method is lower than the non-hierarchical method.

Table 2 (a) P-values considering error of the non-hierarchical and the hierarchical methods with different tolerance parameters. (b) P-values considering cost of the non-hierarchical and the hierarchical methods with different tolerance parameters

a	P-values	b	P-values
$\tau = 0.8$	0.013414	$\tau = 0.8$	$\ll 0.001$
$\tau = 0.6$	0.046626	$\tau = 0.6$	$\ll 0.001$
$\tau = 0.4$	0.635566	$\tau = 0.4$	$\ll 0.001$
$\tau = 0.2$	0.001309	$\tau = 0.2$	$\ll 0.001$
$\tau = 0.1$	$\ll 0.001$	$\tau = 0.1$	$\ll 0.001$
$\tau = 0.0$	$\ll 0.001$	$\tau = 0.0$	$\ll 0.001$

3.2 Actual Data

In this experiment, ten datasets are selected from the UCI database repository [2] based on the number of instances and features from the regression section. Table 3 summarizes these datasets and their features. For these datasets, we partition each into two halves for training and testing, and treat the individual features as individual sensors. Each experiment in this section was repeated 30 times.

In order to determine the accuracy of each sensor s_i in predicting $g^{(t)}$, the value of $v_{r_i}^{(g)}$ is calculated for each available sensor of each dataset, using linear regression. The greater $v_{r_i}^{(g)}$ is, the better that sensor can predict $g^{(t)}$. Table 4 summarizes the values of $v_{r_i}^{(g)}$ for all of the sensors of the Auto MPG dataset, as an example. We define the cost of each sensor in these datasets as $v_{r_i}^{(g)}$ (Table 5).

Table 3 Used UCI datasets

DS no.	DS name	No. of instances	No. of sensors	$g^{(t)}$ Average
DS_1	Auto MPG	398	7	23.51457
DS_2	Housing	506	13	22.53281
DS_3	Forest fires	517	12	0.031663
DS_4	Energy efficiency	768	8	22.3072
DS_5	Concrete compressive strength	1030	8	35.81796
DS_6	Solar flare	1389	9	0.300188
DS_7	Airfoil self-noise	1503	5	124.8359
DS_8	SkilCraft1 master table dataset	3395	19	4.184094
DS_9	Wine quality	4898	11	5.877909
DS_{10}	Parkinson's telemonitoring	5875	17	29.01894

Table 4 Value of $v_{r_i}^{(g)}$ for all of the sensors of Auto MPG dataset

DS no.	s_1	s_2	s_3	s_4	s_5	s_6	s_7
Auto MPG	0.1766	0.3175	0.3356	0.5951	0.6012	0.6467	0.6918

Table 5 Minimum and maximum amount of variance a sensor accounts for and the order of their difference $\dfrac{\max v_{r_i}^{(g)}}{\min v_{r_i}^{(g)}}$, in each dataset

DS no.	$\min v_{r_i}^{(g)}$	$\max v_{r_i}^{(g)}$	Difference ratio
DS_1	0.1766	0.6918	3.92
DS_2	0.0307	0.5441	17.72
DS_3	0.0002	0.2578	1289
DS_4	0.0076	0.7911	104.10
DS_5	0.0112	0.2478	22.13
DS_6	0.000	0.096	96
DS_7	0.0157	0.1527	9.73
DS_8	0.0005	0.4542	908.40
DS_9	0.0001	0.1897	1897
DS_{10}	0.0037	0.0263	7.11

Non-hierarchical setup The population size is 200 and for each dataset with k features, it is trained for $200 \times k$ generations.

Hierarchical setup The population size for each layer is 200. Similar to synthesized data experiments, in order to equalize search effort in both methods, each layer was trained for 200 generations. After training, a subset of the non-dominated solutions with least error are selected and organized in the corresponding layer. The cardinality of this subset is 2 % of the population size. This experiment was conducted for tolerance parameter $\tau = 0.1$. This value is selected based on the results in Sect. 3.1 and will be discussed in more detail in Sect. 5.1.

Results on Actual Data We now consider average error and cost of the different modeling approaches on actual data obtained from UCI data repository.

Average error The average error for the non-hierarchical and the hierarchical methods are $e_{t_j}^{\text{test}}$ and $e_{L_i}^{\text{test}}$ respectively, where t_j is the final selected solution in the non-hierarchical method and L_i is the last layer reached during the sampling in the hierarchical method. Table 6 summarizes the average error of both methods on all of the datasets as a percentage of error. It can be seen that for 6 datasets, the average error of the hierarchical method is higher than the average error of the non-hierarchical

Table 6 Average error (L) and cost (R) percentages and the corresponding P-values for the hierarchical and the non-hierarchical methods

DS no.	NH: error %	H: error %	P-value	NH: cost %	H: cost %	P-value
DS_1	20.85	25.81	$\ll 0.001$	38.89	12.33	0.022
DS_2	25.90	28.93	$\ll 0.001$	23.18	1.26	$\ll 0.001$
DS_3	126.49	202.12	0.565	6.83	15.81	$\ll 0.001$
DS_4	29.19	36.70	$\ll 0.001$	32.90	4.18	$\ll 0.001$
DS_5	35.29	39.68	0.393	53.63	28.63	0.004
DS_6	110.63	111.08	0.223	0.00	0.98	0.040
DS_7	0.00	0.00	0.082	11.58	7.35	0.005
DS_8	37.59	28.65	0.194	2.55	0.00	$\ll 0.001$
DS_9	10.79	10.67	0.197	0.02	0.00	$\ll 0.001$
DS_{10}	32.11	29.88	0.423	20.62	3.88	0.009

method. However, the p-value for the two-tailed t-test shows that for 3 datasets, this difference is not significant. There are three cases where the difference is significant i.e., DS_1, DS_2 and DS_4.

Average cost Similar to Sect. 3.1, the average cost is represented as the percentage of the maximum possible cost. Table 6 summarizes the percentage of the average cost each method uses for prediction. The cost of the hierarchical method is significantly lower in all cases except for DS_3 and DS_6.

4   Adapting to Dynamic Budgets

In remote sensor deployments, the cost associated with sensor sampling may have an effect on the *budget* available. Budget fluctuations can be due to various reasons, depending on the network and the particular definition of the budget. For example, if the budget is defined to be the capacity of a solar rechargeable battery powering the sensor system, the budget may increase on a sunny day, regardless of sampling frequencies, and may decrease on a cloudy day or at night due to sensor usage and battery draw-down. In fact, battery power levels in systems with solar recharging often exhibit a consistently diurnal pattern.

Since it is possible for budgets to fluctuate, a cost-sensitive approach to sensor sampling will ideally *adapt* to changing budget levels, in order to extend deployment lifetimes. In particular, as budgets decrease, models should be biased more towards use of less-costly sensors, to preserve the existing budget and prevent using the entire budget. In the case where the budget is taken to be the battery power level, complete use of the budget corresponds to complete battery drawdown—a potentially catastrophic situation that generally should be avoided.

In this section we reconsider the hierarchical and non-hierarchical methods described previously, with modifications to adapt to fluctuating budgets. In the case of the hierarchical method, we adapt models by allowing the threshold τ to be dynamically tuned in proportion to the remaining budget. In the case of the non-hierarchical method, we add the remaining budget as an input parameter to training and testing. Our main goal is to explore the relative performance of models generated by these respective methods. All of the material for replicating the work described here is also available online [1].

A crucial element of this investigation is the concept of *noise*. Active sensors typically have thresholds for reliable use, and as power levels drop near and then below these levels, sensor noise increases. We observe that this phenomena actually benefits adaptation to budget levels in model training, since increased noise increases error and hence discourages sampling. We consider in particular the scenario where more expensive sensors experience more noise as sensor levels drop—this scenario has an empirical basis in the experience of the authors [3], and has the added benefit (as we will show) of greater bias towards less expensive sensors as budget levels decrease.

Summary of Training and Testing Regimes To encourage adaptation to fluctuating budgets, during training each model is exposed to two different environments: one with a "high" budget and the other with a "low" budget, relative to a posited lower threshold for sensor inputs. Note that only the non-hierarchical models will use the budget level as an input parameter, but predictions of models generated by both methods experience noise proportional to the budget level and the cost of sensors used in the prediction. The optimization objectives for both the hierarchical and non-hierarchical regimes remain the same as in the preceding experiments.

After training, the resultant models are tested on four conditions: each condition takes one of two different initial budgets—high and low—and one of two different budget behaviors: one that stays constant until drawn down by sensor use, and another that has an underlying sinusoidal pattern that simulates diurnal replenishing from solar recharging. Models that exhibit low error and cost in all four situations are considered most desirable.

4.1 Problem Formalization

Let $B^{(t)}$ be a real number defining the amount of the available budget at time t if none of the sensors is polled from the first sampling $\mathbf{S}^{(1)}$ to the last sampling before now $\mathbf{S}^{(t-1)}$. Then, $\mathbf{B}$ is the vector of the available budget for all of the sampling times without any sensor being polled.

Let $\mathbf{B}_H$ and $\mathbf{B}_L$ be the vectors reporting the currently available budget if none of the sensors are polled, where the budget was initially 'high' or 'low'. We denote individual budget values in these vectors as $B_H^{(t)}$ or $B_L^{(t)}$, respectively. The initial budget is considered to be high if the model has enough of a budget to poll two thirds of the available sensors for each sampling,

$$B_H^{(1)} \geq |d^{train}|\left(\frac{2}{3}\sum_{i=1}^{|S|} c_i\right)$$

The average cost of the hierarchical and the non-hierarchical models reported in Sect. 3 are all less than $\mathbf{B}_H$. That is the reason we believe this is a good threshold for the high budget level. If the budget is not enough to poll at least one third of the available sensors for each sampling, then it is considered to be low:

$$B_L^{(1)} \leq |d^{train}|\left(\frac{1}{3}\sum_{i=1}^{|S|} c_i\right)$$

Let $c_{t_j,\mathbf{B}}^{(t)}$ denote the cost of evaluating solution tree t_j at time t, considering the available budget $\mathbf{B}$, which could in turn be drawn from $\mathbf{B}_L$ or $\mathbf{B}_H$. Note that $c_{t_j,\mathbf{B}}^{(t)}$ depends on the particular solution tree t_j and the sensors used by that model, as explained in Sect. 4. This cost should be deducted from the currently available budget.

Let $R_{t_j,\mathbf{B}}^{(t)}$ then denote the amount of remaining budget at time t, considering budget $\mathbf{B}$ for each solution tree t_j. Then, $R_{t_j,\mathbf{B}}^{(t)}$ can be defined as

$$R_{t_j,\mathbf{B}_b}^{(t)} = B^{(t)} - \sum_{l=1}^{t-1} c_{t_j,\mathbf{B}_b}^{(l)}, b \in \{H, L, \epsilon\}$$

We define $R_{t_j,\mathbf{B}_H}^{(t)}$ and $R_{t_j,\mathbf{B}_L}^{(t)}$ as the remaining budgets when the budget $\mathbf{B}$ being used is either the high budget $\mathbf{B}_H$ or the low budget $\mathbf{B}_L$.

It is notable that, as explained in Sect. 4, the accuracy of sensors is affected by the level of the remaining budget. By decreasing the amount of the available budget, the error of a sensor S_i and the amount of noise in its reading r_i will increase. Moreover, we consider a noise model in which different sensors may be affected by a reading differently: we assume that less expensive sensors become less noisy as the level of the remaining budget drops, since they are less costly and thus reduce the budget less than expensive sensors. This behaviour is modelled as follows, where $U(\min(r_i), \max(r_i))$ is a uniform random number from the r_i domain:

$$r_i'^{(t)}(t_j, \mathbf{B}) = \left(\frac{R_{t_j,\mathbf{B}}^{(t)}}{B^{(t)}}\right)^{c_i} r_i^{(t)} + \left(1 - \left(\frac{R_{t_j,\mathbf{B}}^{(t)}}{B^{(t)}}\right)^{c_i}\right) U(\min(r_i), \max(r_i))$$

In this manner, by decreasing the available budget, the accuracy of the sensor readings will decrease and the noise in the sensor readings will increase. This effect is proportionate to the cost. Figure 4 shows the rate of noise as the available budget level drops for three sensors in the synthesized dataset introduced in Sect. 3.1. As

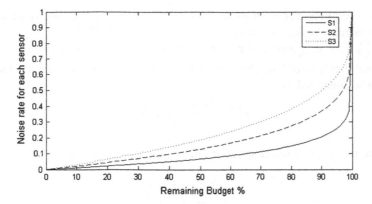

Fig. 4 The noise rate for different sensors with different cost from synthesized dataset

shown there, the cheapest sensor S_1 becomes less noisy with decreasing budget, whereas the most expensive sensor S_3 suffers a greater noise increase as the budget decreases.

This noise model should encourage the selection of models that make use of less expensive sensors for predictions for two reasons. First, less expensive sensors become less noisy when the available budget level drops, compared to more expensive sensors. Second, using less expensive sensors keeps the cost of each prediction low. Since the prediction cost has to be paid for from the available budget, low cost models cause a slower decrease in the budget and thus retain more accurate sensor readings.

When the accuracy of a sensor s_i is affected by the available budget, then the accuracy of the solution tree t_j that makes use of s_i also suffers. Let $e^{(t)}_{t_j, \mathbf{B}}$ denote the error of solution tree t_j at time t, and where the available budget is $\mathbf{B}$.

4.2 Methods

In order for the models to adapt to a changing budget, both the hierarchical and non-hierarchical methods described in Sect. 2 should be able to alter their prediction strategies, given the current budget.

Non-hierarchical GP In order to enable the non-hierarchical models to modify their prediction strategy given the current budget, we include the currently remaining budget $R^{(t)}_{t_j, \mathbf{B}}$ as an additional 'sensor' that can be incorporated into solution trees during model training. This is realistic since budget information such as power level data is frequently accessible in WSN systems. If this sensor is incorporated into a model, it can read the level of the remaining budget at no cost, when it is needed.

The solution trees are trained in the same way described in Sect. 2.3, except for how the trees are evaluated. Each solution tree t_j is evaluated twice, once using $\mathbf{B}_H$ and once using $\mathbf{B}_L$, to encourage model robustness. These two budget distributions are considered to be flat, without any budget harvesting: that is, the overall budget does not increase or decrease over time if no sensors are polled.

Let $e^{(t)}_{t_j,\mathbf{B}_H}$ and $e^{(t)}_{t_j,\mathbf{B}_L}$ denote the error of a solution tree t_j at time t when the budget level was either high or low. The error of a solution tree t_j at time t, denoted as $e^{(t)}_{t_j}$, can then be computed as

$$e^{(t)}_{t_j} = \frac{e^{(t)}_{t_j,\mathbf{B}_H} + e^{(t)}_{t_j,\mathbf{B}_L}}{2} \tag{4}$$

Similarly, the cost of a solution tree at time t, $c^{(t)}_{t_j}$, is computed as

$$c^{(t)}_{t_j} = \frac{c^{(t)}_{t_j,\mathbf{B}_H} + c^{(t)}_{t_j,\mathbf{B}_L}}{2} \tag{5}$$

With this formulation, the average error of a solution tree $e^{\text{train}}_{t_j}$, and the overall cost of a tree $e^{\text{train}}_{t_j}$, can be calculated based on Eqs. 1 and 2 given in Sect. 2.2.

Hierarchical GP The hierarchical method is trained the same as described in Sect. 2.4 except that, like the non-hierarchical method, models in the hierarchical method are trained on both $\mathbf{B}_H$ and $\mathbf{B}_L$, and their respective costs and errors are computed as the average cost and error incurred in these two budget regimes (Eqs. 4 and 5). In this manner, each layer L_i consists of models with high robustness over different budget distributions.

In Sect. 2.4 the tolerance parameter τ is statically defined and does not change during model execution. Here however, as the budget is dynamic, the tolerance parameter should change accordingly. Therefore, τ is defined as

$$\tau' = 1 - \frac{R^{(t)}_{t_j,\mathbf{B}}}{B^{(t)}}$$

This balances which layer of the model hierarchy provides predictions, given the currently remaining budget. When the remaining budget is high, the threshold for disagreement between models of a given layer is low, so predictions tend to be drawn from higher layers which have high accuracy. A low remaining budget means that the threshold for disagreement between models of a given layer is high, thus relegating predictions to lower levels of the model hierarchy. This has the effect of causing the overall hierarchical model to become increasingly conservative in its use of sensors as the budget decreases.

Substituting the tolerance parameter τ with this new τ' in Eq. 3 given in Sect. 2.4 thus results in a new condition for switching between layers:

$$p^{(t)} = \begin{cases} p^{(t)}_{L_i} & \text{if } v^{\text{train}}_i > \dfrac{R^{(t)}_{t_j,\mathbf{B}}}{B^{(t)}} \cdot v^{\text{test}(t)}_i \\ p^{(t)}_{L_{i+1}} & \text{otherwise} \end{cases}$$

4.3 Results

These altered methods for training models were evaluated against two sets of data: a synthesized dataset and ten actual datasets, as described in Sect. 3. This section summarizes the results from training with these datasets.

Each model is trained with given fixed budgets $\mathbf{B}_H$ and $\mathbf{B}_L$. These budget distributions are defined such that for all times t we have:

$$B_b^{(t)} = |d^{train}|(\alpha \sum_{i=1}^{|S|} c_i) \quad \text{where} \begin{cases} \alpha = 1/3 & \text{if } b = L \\ \alpha = 2/3 & \text{if } b = H. \end{cases}$$

After training models for each dataset, they are tested on four budget distributions $\mathbf{B}_H$, $\mathbf{B}_L$, $\mathbf{B}_{H,\text{sin}}$ and $\mathbf{B}_{L,\text{sin}}$. Budget distributions $\mathbf{B}_H$ and $\mathbf{B}_L$ are flat and the same as training datasets. Budget distributions $\mathbf{B}_{H,\text{sin}}$ and $\mathbf{B}_{L,\text{sin}}$ were constructed to simulate diurnal replenishing of solar powered sensors. This was accomplished by adding a sinusoidal pattern to $\mathbf{B}_H$ and $\mathbf{B}_L$. The amplitude of the sine wave is set to 2 % of the high budget level. We let $B_{b,\text{sin}}^{(t)}$ for $b \in \{H, L\}$ denote the budget value at time t in a given sinusoidal distribution $\mathbf{B}_{H,\text{sin}}$ and $\mathbf{B}_{L,\text{sin}}$. We define the latter such that for any time t we have:

$$B_{b,\text{sin}}^{(t)} = B_b^{(t)} + \sin\left(\frac{t}{\lceil B_H^{(t)} * 0.02 \rceil}\right) \quad b \in \{\epsilon, H, L\}$$

Synthesized Data In these experiments, the synthesized data described in Sect. 3.1 is used to evaluate the proposed methods. The training and testing datasets are the same, except that budget is also included as an extra sensor. The training and testing datasets both contain 150 samples, and each experiment is repeated 30 times for each budget distribution.

Non-hierarchical model training The population size is set to 100, and models are trained for 600 generations, or more precisely 200 generations multiplied by the number of available sensors, 3 in this case. The budget sensor is available to the non-hierarchical model. The rest of the settings are the same as described in Sect. 3.1. After training, the selected model is tested on all four different budget distributions.

Hierarchical model training The population size for each layer is set to 100, and each of the three layers are trained for 200 generations. For training each layer L_i, the corresponding sensor S_i and the less expensive ones $\{S_j | j < i\}$ are provided as input. Models do not have access to the remaining budget level as an extra feature in training. The rest of model training is as described in Sect. 3.1. During testing, the dynamic tolerance parameter τ' is used.

Results on Synthesized Data If the non-hierarchical model selected for testing is t_j, the average cost of the non-hierarchical model is equal to $c_{t_j,\mathbf{B}}^{(t)}$ averaged over the testing dataset d^{test}. The average error of the non-hierarchical model is also equal to the error of each sampling $e_{t_j,\mathbf{B}}^{(t)}$ averaged over the testing dataset d^{test}. The budget

Fig. 5 Average error (**a**) and cost (**b**) on the test data for the non-hierarchical and the hierarchical models with different budget distributions. Statistical significance of these results are reported in Table 7

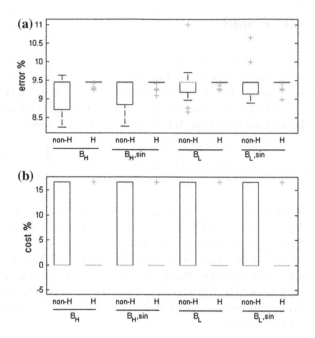

distribution **B** could be one of the four given distributions. The sensor readings are denoted as r_i', which reflect the noise considering the remaining budget level as described in Sect. 4.1.

Let $L^{(t)}$ denote the layer in the hierarchical method that the prediction is drawn from at time t. Then, the average cost of the hierarchical model at time t is equal to $c_{L^{(t)},\mathbf{B}}^{(t)}$, averaged over the testing dataset d^{test}. The average error of the hierarchical method is equal to the error of layer $L^{(t)}$ averaged over the testing dataset. The error of a layer L_i is equal to

$$e_{t_{i,j},\mathbf{B}}^{(t)} = \frac{1}{|L_i|} \sum_{j=1}^{|L_i|} e_{i,j}^{(t)}$$

where $|L_i|$ defines the number of solution trees in layer L_i and $e_{i,j}^{(t)}$ is the error of the jth solution in layer L_i when sensor readings are r_i'.

As can be seen in Fig. 5, the error rates between both methods over all four budget distributions are not significantly different (at most 0.32 %). Table 7a makes clear that there are no statistically significant differences in errors across methods when the initial budget is low. Figure 5 also shows that the average cost of hierarchical models is significantly lower than non-hierarchical models (at most 7.3 %). Table 7b summarizes the statistical significance of these differences.

Table 7 (a) P-values considering error of the non-hierarchical and the hierarchical methods with different budget distribution. (b) p-values considering cost of the non-hierarchical and the hierarchical methods with different budget distribution

(a)	p-values	(b)	p-values
$\mathbf{B}_H$	≪0.001	$\mathbf{B}_H$	≪0.001
$\mathbf{B}_{H,\sin}$	≪0.001	$\mathbf{B}_{H,\sin}$	≪0.001
$\mathbf{B}_L$	0.612857	$\mathbf{B}_L$	≪0.001
$\mathbf{B}_{L,\sin}$	0.590440	$\mathbf{B}_{L,\sin}$	≪0.001

Actual Data In these experiments, datasets from the UCI repository, as described in Sect. 3.2, are used to evaluate the two new proposed methods. As in Sect. 3.2, each dataset is divided into two equal training and testing portions. The budget feature is also included in the datasets when training the non-hierarchical models. Each experiment is repeated 30 times. For each iteration, a model is selected and tested on the test data four times, each time with a different budget distribution.

Non-hierarchical model training The population size is set to 100, and for each datasets with k available sensors, models are trained for $200 \times k$ generations. The non-hierarchical models have access to the $\mathbf{B}$ feature during training. The rest of the settings are the same as they are reported in Sect. 3.2. After training, the selected model is tested on all four different budget distributions.

Hierarchical model training The population size for each layer is set to 100, and for each dataset, each of the layers are trained for 200 generations. Models do not have access to the remaining budget level as an extra feature in training. The rest of model training is as described in Sect. 3.2. During testing, the dynamic tolerance parameter τ' is used.

Table 8 Error percentages of the methods on actual data considering different budget distributions

Datasets	$\mathbf{B}_H$		$\mathbf{B}_{H,\sin}$		$\mathbf{B}_L$		$\mathbf{B}_{L,\sin}$	
	non-H	H	non-H	H	non-H	H	non-H	H
DS_1	**22.03**	27.95	**21.97**	27.94	**23.29**	29.17	**23.30**	29.41
DS_2	32.14	**31.13**	32.19	**31.08**	32.51	**31.21**	32.54	**31.16**
DS_3	**99.61**	99.67	**99.61**	99.68	**99.61**	99.66	**99.61**	99.67
DS_4	**39.63**	43.18	**39.71**	43.19	43.79	**43.03**	43.72	**43.01**
DS_5	**40.59**	49.50	**40.64**	49.48	**42.26**	49.51	**42.32**	49.49
DS_6	**99.81**	101.2	**99.91**	101.2	**99.94**	101.2	**99.91**	101.2
DS_7	0.00	0.00	0.00	0.00	0.00	0.00	0.00	0.00
DS_8	**36.55**	42.50	**36.55**	42.45	**36.89**	42.49	**36.93**	42.50
DS_9	11.60	**11.57**	11.60	**11.57**	11.60	**11.57**	11.60	**11.57**
DS_{10}	37.83	**37.78**	37.82	**37.77**	**37.83**	37.91	**37.83**	37.84

The smallest of the non-hierarchical and the hierarchical methods are indicated in bold

Table 9 P-values for actual data comparing the error of the hierarchical and non-hierarchical method considering different budget distributions

Datasets	$\mathbf{B}_H$	$\mathbf{B}_{H,\sin}$	$\mathbf{B}_L$	$\mathbf{B}_{L,\sin}$
DS_1	≪0.001	≪0.001	≪0.001	≪0.001
DS_2	0.005	0.003	0.012	0.008
DS_3	0.310	0.314	0.312	0.313
DS_4	0.002	0.002	0.486	0.511
DS_5	≪0.001	≪0.001	≪0.001	≪0.001
DS_6	0.003	0.003	0.004	0.007
DS_7	1.00	1.00	1.00	1.00
DS_8	≪0.001	≪0.001	≪0.001	≪0.001
DS_9	0.680	0.689	0.709	0.695
DS_{10}	0.950	0.937	0.899	0.990

Results on Actual Data Table 8 reports the average prediction errors for all of the actual datasets, for both the hierarchical and non-hierarchical methods. The statistical significance of the difference in errors between these two methods is reported in Table 9. The average cost of models trained on the actual datasets for the hierarchical and non-hierarchical methods can be seen in Table 10. Table 11 reports the statistical signficance of the cost differences between the two methods.

Table 10 Cost percentages of the methods on actual data considering different budget distributions

Datasets	$\mathbf{B}_H$		$\mathbf{B}_{H,\sin}$		$\mathbf{B}_L$		$\mathbf{B}_{L,\sin}$	
	non-H	H	non-H	H	non-H	H	non-H	H
DS_1	15.23	**8.84**	15.23	**8.80**	15.23	**8.30**	15.23	**8.27**
DS_2	9.08	**1.84**	9.09	**1.85**	9.08	**1.74**	9.09	**1.76**
DS_3	0.031	**0.018**	0.030	**0.017**	0.031	**0.017**	0.031	**0.017**
DS_4	30.21	**0.631**	30.23	**0.631**	30.06	**0.587**	30.04	**0.597**
DS_5	46.93	**3.17**	46.93	**3.18**	46.93	**3.06**	46.93	**3.04**
DS_6	0.412	**0.001**	0.412	**0.001**	0.412	**0.001**	0.412	**0.001**
DS_7	**1.02**	2.73	**1.02**	2.73	**1.02**	2.65	**1.02**	2.67
DS_8	0.176	**0.00**	0.176	**0.00**	0.176	**0.00**	0.176	**0.00**
DS_9	0.076	**0.00**	0.076	**0.00**	0.076	**0.00**	0.076	**0.00**
DS_{10}	8.00	**1.39**	8.00	**2.21**	8.00	**2.06**	7.99	**2.06**

The smallest of the non-hierarchical and the hierarchical methods are indicated in bold

Table 11 P-values for actual data comparing the cost of the hierarchical and non-hierarchical method considering different budget distributions

Datasets	$\mathbf{B}_H$	$\mathbf{B}_{H,\sin}$	$\mathbf{B}_L$	$\mathbf{B}_{L,\sin}$
DS_1	$\ll 0.001$	$\ll 0.001$	$\ll 0.001$	$\ll 0.001$
DS_2	$\ll 0.001$	$\ll 0.001$	$\ll 0.001$	$\ll 0.00086$
DS_3	$\ll 0.001$	$\ll 0.001$	$\ll 0.001$	$\ll 0.001$
DS_4	$\ll 0.001$	$\ll 0.001$	$\ll 0.001$	$\ll 0.001$
DS_5	$\ll 0.001$	$\ll 0.001$	$\ll 0.001$	$\ll 0.001$
DS_6	0.024	0.024	0.025	0.024
DS_7	0.513	0.515	0.618	0.585
DS_8	$\ll 0.001$	$\ll 0.001$	$\ll 0.001$	$\ll 0.001$
DS_9	0.164	0.164	0.164	0.164
DS_{10}	0.019	0.019	0.016	0.016

5 Discussion

In this section we reflect on the reason for and meaning of our quantitative results described in Sects. 3 (basic results) and 4.3 (results with a dynamic budget). Overall, our experimental results show that in any case, sampling costs of models generated by the hierarchical method are significantly lower than models generated by the non-hierarchical method. Non-hierarchical modes use more expensive sensors with higher frequency. Results also show that hierarchical models achieve similar error rates as those incurred by non-hierarchical models as datasets grow larger, though non-hierarchical models do achieve lower error for small datasets especially when a dynamic budget is considered. Also notable is that results in Sect. 4.3 suggest that the hierarchical method obtains models that are more effectively sensitive to noise than models generated by the non-hierarchical method, when the budget level shrinks from high to low.

5.1 Basic Results with a Static Budget

The results presented in Sect. 3 suggest that the hierarchical method is better at balancing cost and accuracy than the non-hierarchical approach. We believe this is because meaningful sensor control conditions for managing cost are complex and require considerable computational effort to be discovered. Using hand-tuned prediction uncertainty to drive sensor control is more effective. As mentioned in Sect. 2.2, in these experiments a basic genetic programming approach was deployed.

We anticipate that if we were to use a more powerful underlying GP approach, the error of both hierarchical and non-hierarchical models would be reduced.

Synthesized Data The hierarchical method achieved significantly better accuracy and significantly lower cost than the non-hierarchical using synthesized data.

Average error As can be seen in Fig. 3, the hierarchical method achieved significantly better accuracy than the non-hierarchical method for $\tau < 0.4$. In general, results show that higher tolerance allows the algorithm to accept more uncertainty in the prediction and rely on less expensive sensors which are less accurate. This avoids the use of more expensive sensors, but causes average error to rise. A tolerance of $\tau < 0.4$ is apparently the threshold where average error in the hierarchical method exceeds that of the non-hierarchical method, when analysing results for different values of tolerance included in this study.

Average cost Results reported in Fig. 3 show that the hierarchical method significantly outperforms the non-hierarchical method with regard to cost on this dataset, even when tolerance is low. This suggests that the use of variance in ensemble predictions to serve as a proxy for prediction uncertainty is not easy to learn, and serves as a good mechanism for control. Results suggest that $\tau = 0.1$ is a "sweet spot" for balancing cost and accuracy, though the value could be increased or decreased if greater frugality or accuracy were needed, respectively.

Actual Data For testing on actual data, we fixed $\tau = 0.1$ due to results on synthetic data demonstrating a good balance between cost and accuracy with this tolerance level.

Average error Table 6 shows that the average error of the hierarchical and the non-hierarchical methods were not significantly different, except for datasets DS_1, DS_2 and DS_4 where the latter method achieves better prediction accuracy. This is probably due to the characteristics of these datasets, where the difference between the least prediction variances $v_{r_i}^{(g)}$s and the greatest ones is large. The majority of sensors in these datasets are not informative but have low costs and the remaining sensors are more informative but come with higher costs. Thus, lower levels of the hierarchy "struggle" compared to upper ones in terms of accuracy. Nevertheless, accuracy rate with the hierarchical method is still competitive even in these cases, and cost reduction is significant. Also, it can be seen that as the size of the datasets grows, the difference between the error rate of the non-hierarchical and the hierarchical methods decreases, and in the three largest datasets the hierarchical method also achieves lower error rates.

Average cost The hierarchical method achieved significantly lower cost than the non-hierarchical method on all of the real world datasets, as shown in Table 6, except for DS_3 and DS_6. As represented in Table 5, in these two datasets, just a small subset of sensors are relatively informative. Since the tolerance parameter for the hierarchical method is low, the hierarchical method employs more informative sensors. Taken together, results shown in Table 6 clearly indicates an advantage of the hierarchical method for balancing cost and accuracy.

5.2   Results with a Dynamic Budget

Now we consider the results provided in Sect. 4.3 on synthesized data and ten actual datasets when a possibly dynamic budget is taken in to account. The results obtained from the experiments in Sect. 4.3 suggest that the hierarchical method is more successful in balancing cost and prediction accuracy compared to the non-hierarchical method as the number of observations in the dataset grows. The hierarchical method produces much less costly models, which results in less noise accumulating on the sensors. This conservation can be crucial to reduce system down time if longer time periods are required to replenish the budget.

It also can be seen in Table 10 that the hierarchical method produces models that adapt their sensor sampling strategy based on the current budget, since they reduce their costs when the budget level goes from high to low. In contrast, models produced by the non-hierarchical method do not change their cost when the models are presented with the low budget level. The reason why the non-hierarchical method does not make use of the remaining budget to change its behaviour is at the moment unclear. Models produced by the hierarchical method incur lower cost when the budget level is low than when the budget level is high. The dynamic tolerance parameter employed in the hierarchical method successfully balances the cost of the hierarchical method to the remaining budget considering the results reported in Fig. 5 and Table 10.

Synthesized Data The results using synthesized data (Fig. 5) demonstrate that the hierarchical method adapts to the changing budget better than the non-hierarchical method. The hierarchical models obtain about the same prediction accuracy as the non-hierarchical models, but with significantly lower cost.

Average error As can be seen in Fig. 5, the difference between the error rate of the hierarchical and non-hierarchical methods is low. The p-values reported in Table 7a suggest that this difference is insignificant when the budget level is low. When the hierarchical method must work within the confines of a low budget, it produces models that only infrequently poll high-cost sensors. In this manner, the hierarchical models keep the overall cost of prediction low, which results in a higher remaining budget for the remainder of the period during which predictions are requested. Keeping the budget high in turn results in sensor readings with higher accuracy.

The non-hierarchical models however maintain high accuracy by polling the more accurate sensors more frequently, which incurs a higher cost. This approach eventually causes prediction accuracy to suffer, since it increases the noise in sensor readings as the budget decreases. As can be seen in Table 7a, when the budget is low, the non-hierarchical models are not able to maintain their superior accuracy rates.

Average cost Figure 5 shows that the hierarchical method generates significantly lower cost models compared to the non-hierarchical method, for all of the budget distributions considered. The hierarchical method keeps cost low in two ways. First, the model hierarchy is constrained by design in the sensors it samples, depending on the hierarchy level. Second, the certainty threshold can be tuned to become more restrictive as the budget drops.

If the budget is low, then the dynamic tolerance parameter forces the hierarchical model to tolerate more uncertainty in its predictions. In contrast, the non-hierarchical method generates models that tend to use more expensive sensors more frequently. The results shown in Table 7b support this claim. As in the basic setting with a static budget, the non-hierarchical method has difficulties discovering the proper conditions for sampling various sensors, that can be manually tuned into the threshold parameter for the hierarchical method.

Actual Data With actual data, the hierarchical models are able to adapt to budget fluctuations better than the non-hierarchical models, but the non-hierarchical models achieve higher accuracy on smaller datasets.

Average error Table 8 shows that the non-hierarchical method achieves better error rates compared to the hierarchical method on datasets for which most of the sensors are non informative, but a few are with very high cost. For example, as shown in Table 5, in dataset DS_8 with 19 sensors, the difference in accuracy and cost between the most informative and the least informative sensors is on the order of 10^3. The most informative sensor is able to explain 45.42 % of the output variance while the least informative sensor explains just 0.05 % of the output variance. The difference in accuracy and cost between the other sensors and the most informative sensor are almost the same, except for the four most informative sensors. In this case, the non-hierarchical method uses the informative sensors in order to achieve high accuracy, whereas the hierarchical method tries to find a model with less cost. The order of difference for the most informative and the least informative sensors in DS_3 and DS_9 is high, but this difference order reduces for the other sensors in those datasets and also the most informative sensor in these datasets are not that informative (25.89 % and 18.97 % respectively). For the rest of the datasets, the order of difference is not that high compared to DS_8.

Also, in the non-hierarchical method, a model and its descendants could have been refined through the entire training period, whereas in the hierarchical method the training effort is distributed among the hieararchy layers. This means that individual model lineages have much less time to be improved, compared to non-hierarchical model lineages. Even so, Tables 8 and 10 show that when the budget level drops from high to low, the cost of the hierarchical models drops further than the drop observed in the non-hierarchical models. In this way the budget is better conserved during deployment and sensor noise is ameliorated.

As can be seen in Table 8, the error rates of the non-hierarchical models grow more than the error rates of the hierarchical models when the budget level decreases. Moreover, as the sizes of the datasets grow (from DS_1 to DS_{10}), the differences between the error rates of the hierarchical and non-hierarchical models decrease. For the two largest data sets (DS_9 and DS_{10}), the average error rate of the hierarchical and the non-hierarchical models are not statistically significantly different.

Average cost As can be seen in Table 10, the hierarchical models always achieve lower cost compared to the non-hierarchical models, except for DS_7. Data set DS_7 has five sensors which all explain a small amount of the outcome variance. Hierarchical models attempt to predict the outcome using these sensor readings, but the non-hierarchical models rarely employ any of those sensors since they have so little predictive value. Instead the hierarchical models use the training effort to find a constant that predicts the outcome, at no cost.

Otherwise, as seen in Table 10, hierarchical models generally reduce cost more than non-hierarchical models when initial costs go from high to low, as is the case for synthesized data and for the same reasons (or so we hypothesize).

6 Conclusion and Future Work

Wireless sensor networks often face a trade-off between measurement accuracy and the cost of sensor sampling. In networks supporting multiple sensor types, it is therefore desirable to develop cost-sensitive control algorithms that sample more expensive sensors only when necessary. In this chapter, a hierarchical method is proposed where GP solutions are sorted in a hierarchy of layers based on the cost of the sensors they use. Switching to the next more expensive layer takes place only if the prediction variance indicates uncertainty at lower layers. We compare this method to a non-hierarchical GP method where cost is treated as an additional optimization objective in fitness selection. In experiments using a synthesized dataset and ten real datasets, the hierarchical method is shown to have significantly lower prediction costs than the non-hierarchical method. As the datasets grow larger and more complex, competitive and sometimes lower error rates are achieved by the hierarchical method. In a second set of experiments, we consider a more sophisticated setting where the current budget level (e.g. power levels) is available as a sensor reading, and lower budgets have a direct impact on sensor accuracy. The non-hierarchical method in this case uses the remaining budget level in order to induce a model that adapts to the budget and sensor noise. In the hierarchical method the remaining budget is used in the decision to switch between layers. The results from experiments show that when the methods are altered to dynamically tune the balance of cost and accuracy based on available energy and budget in the presence of noise, the hierarchical method achieves significantly lower cost. As datasets grow larger, the hierarchical method achieves a competitive error rate as compared to the non-hierarchical method. Future work includes a consideration of methods for online learning to support adaptation of control algorithms to particular deployments, and the application of hierarchical control algorithms in real wireless sensor network deployments.

Acknowledgements This work was supported in part by the NSF awards PECASE-0953837 and INSPIRE-1344227.

References

1. github code public repository. http://git.io/vfmGB. Accessed: 2015-04-18
2. UCI machine learning repository. http://archive.ics.uci.edu/ml/datasets.html. Accessed: 2015-02-03
3. Snowcloud: a complete system for snow hydrology research. ACM (2013)
4. Alippi, C., Anastasi, G., Galperti, C., Mancini, F., Roveri, M.: Adaptive sampling for energy conservation in wireless sensor networks for snow monitoring applications. In: IEEE 4th International Conference on Mobile Adhoc and Sensor Systems, MASS 2007, 8–11 October 2007, Pisa, Italy, pp. 1–6 (2007)
5. Bair, E.H., Davis, R.E., Finnegan, D.C., LeWinter, A.L., Guttmann, E., Dozier, J.: Can we estimate precipitation rate during snowfall using a scanning terrestrial lidar? In: International Snow Science Workshop, pp. 923–929. Anchorage, AK (2012)
6. Brockhoff, D., Zitzler, E.: Are all objectives necessary? On dimensionality reduction in evolutionary multiobjective optimization. In: Parallel Problem Solving from Nature-PPSN IX, pp. 533–542. Springer, Berlin (2006)
7. Buckingham, D., Skalka, C., Bongard, J.: Inductive learning of snowpack distribution models for improved estimation of areal snow water equivalent. J. Hydrol. (2015)
8. Donoho, D.L.: Compressed sensing. IEEE Trans. Inf. Theory $52(4)$, 1289–1306 (2006)
9. Hall, D., Llinas, J.: Multisensor Data Fusion. CRC Press, Boca Raton (2001)
10. Hornby, G.: ALPS: the age-layered population structure for reducing the problem of premature convergence. In: Genetic and Evolutionary Computation Conference, GECCO 2006, Proceedings, Seattle, Washington, USA, July 8–12, 2006, pp. 815–822 (2006)
11. de Jong, E.D., Pollack, J.B.: Multi-objective methods for tree size control. Genet. Program. Evolvable Mach. $4(3)$, 211–233 (2003)
12. Kim, D.: Structural risk minimization on decision trees using an evolutionary multiobjective optimization. In: Genetic Programming, pp. 338–348. Springer, Berlin (2004)
13. Koushanfar, F., Slijepcevic, S., Potkonjak, M., Sangiovanni-Vincentelli, A.: Error-tolerant multimodal sensor fusion. In: IEEE CAS Workshop on Wireless Communication and Networking, pp. 5–6 (2002)
14. Koza, J.R.: Genetic Programming: On the Programming of Computers by means of Natural Selection, vol. 1. MIT Press, Cambridge (1992)
15. Krause, A., Guestrin, C., Gupta, A., Kleinberg, J.: Near-optimal sensor placements: maximizing information while minimizing communication cost. In: Proceedings of the 5th International Conference on Information Processing in Sensor Networks, pp. 2–10. ACM (2006)
16. Maleki, S., Pandharipande, A., Leus, G.: Two-stage spectrum sensing for cognitive radios. In: Proceedings of the IEEE International Conference on Acoustics, Speech, and Signal Processing, ICASSP 2010, 14–19 March 2010, Sheraton Dallas Hotel, Dallas, Texas, USA, pp. 2946–2949 (2010)
17. Malloy, M.L., Nowak, R.D.: Near-optimal adaptive compressed sensing. IEEE Trans. Inf. Theory $60(7)$, 4001–4012 (2014)
18. Martinelli, A.: Vision and IMU data fusion: closed-form solutions for attitude, speed, absolute scale, and bias determination. IEEE Trans. Robot. $28(1)$, 44–60 (2012)
19. Papadimitriou, C., Beck, J.L., Au, S.K.: Entropy-based optimal sensor location for structural model updating. J. Vib. Control $6(5)$, 781–800 (2000)
20. Pohl, C., Genderen, J.V.: Review article multisensor image fusion in remote sensing: concepts, methods and applications. Int. J. Remote Sens. $19(5)$, 823–854 (1998)
21. Ren, H., Rank, D., Merdes, M., Stallkamp, J., Kazanzides, P.: Multisensor data fusion in an integrated tracking system for endoscopic surgery. IEEE Trans. Inf Technol. Biomed. $16(1)$, 106–111 (2012)
22. Schmidt, M., Lipson, H.: Age-fitness pareto optimization. In: Riolo, R., McConaghy, T., Vladislavleva, E. (eds.) Genetic Programming Theory and Practice VIII, Genetic and Evolutionary Computation, vol. 8, pp. 129–146. Springer, New York (2011). doi:10.1007/978-1-4419-7747-2_8

23. Seung, H.S., Opper, M., Sompolinsky, H.: Query by committee. In: Proceedings of the Fifth Annual Workshop on Computational Learning Theory, pp. 287–294. COLT '92, ACM, New York, NY, USA (1992). doi:10.1145/130385.130417
24. Skalka, C., Frolik, J.: Snowcloud: a complete data gathering system for snow hydrology research. In: Real-World Wireless Sensor Networks, pp. 3–14. Springer, Berlin (2014)
25. Smith, D., Singh, S.: Approaches to multisensor data fusion in target tracking: a survey. IEEE Trans. Knowl. Data Eng. 18(12), 1696–1710 (2006)
26. Tabari, H., Marofi, S., Abyaneh, H.Z., Sharifi, M.R.: Comparison of artificial neural network and combined models in estimating spatial distribution of snow depth and snow water equivalent in Samsami basin of Iran. Neural Comput. Appl. 19(4), 625–635 (2010)
27. Tappeiner, U., Tappeiner, G., Aschenwald, J., Tasser, E., Ostendorf, B.: GIS-based modelling of spatial pattern of snow cover duration in an alpine area. Ecol. Model. 138, 265–275 (2001)
28. Waltz, E.L., Buede, D.M.: Data fusion and decision support for command and control. IEEE Trans. Syst. Man Cybern. 16(6), 865–879 (1986)
29. Wang, D., Ahmadi, H., Abdelzaher, T.F., Chenji, H., Stoleru, R., Aggarwal, C.C.: Optimizing quality-of-information in cost-sensitive sensor data fusion. In: Distributed Computing in Sensor Systems, 7th IEEE International Conference and Workshops, DCOSS 2011, Barcelona, Spain, 27–29 June, 2011, Proceedings, pp. 1–8 (2011)
30. Willett, R., Martin, A., Nowak, R.: Backcasting: adaptive sampling for sensor networks. In: Proceedings of the Third International Symposium on Information Processing in Sensor Networks, IPSN 2004, Berkeley, California, USA, April 26–27, 2004, pp. 124–133 (2004)
31. Yick, J., Mukherjee, B., Ghosal, D.: Wireless sensor network survey. Comput. Netw. 52(12), 2292–2330 (2008)
32. Zhao, H.: A multi-objective genetic programming approach to developing pareto optimal decision trees. Decis. Support Syst. 43(3), 809–826 (2007)
33. Zowj, A.Y., Bongard, J.C., Skalka, C.: A genetic programming approach to cost-sensitive control in resource constrained sensor systems. In: Proceedings of the Genetic and Evolutionary Computation Conference, GECCO 2015, Madrid, Spain, July 11–15, 2015, pp. 1295–1302 (2015)

A Study on Performance of Hill Climbing Heuristic Method for Router Placement in Wireless Mesh Networks

Evjola Spaho, Alda Xhafa, Donald Elmazi, Fatos Xhafa and Leonard Barolli

Abstract Wireless Mesh Networks (WMNs), also referred to as a form of wireless ad hoc network are in fact one particular type of Wireless Sensor Networks (WSNs), whose topology can vary from a simple star network to an advanced multi-hop wireless mesh network. The main topological feature in this case is that nodes are organized in a mesh topology, making WMNs a reliable infrastructure through redundancy of multi-hop communications. The main issue of WMNs is to achieve network connectivity and stability as well as Quality of Service (QoS) in terms of user coverage. This problem is very closely related to the family of node placement problems in WMNs, among them, the mesh router mesh nodes placement. In this work we present some optimization problems in WMNs and Hill Climbing (HC) heuristic method for solving mesh router node placement near-optimally. We formulate the optimization problems using bi-objective optimization models. Thus, for the mesh router nodes placement, the bi-objective optimization problem is obtained consisting in the maximization of the size of the giant component in the mesh routers network (for measuring network connectivity) and that of user coverage. Some computational results are presented and discussed for the HC method, which is an effective local search method.

Keywords WMNs · WSNs · Hill Climbing · Heuristic method · Size of giant component · Covered mesh clients

E. Spaho (✉)
Polytechnic University of Tirana, Tirana, Albania
e-mail: espaho@fti.edu.al

A. Xhafa
Universitat Autònoma de Barcelona, Barcelona, Spain
e-mail: aldaxhafa@yahoo.com

D. Elmazi · L. Barolli
Fukuoka Institute of Technology, Fukuoka, Japan
e-mail: donald.elmazi@gmail.com

L. Barolli
e-mail: barolli@fit.ac.jp

F. Xhafa
Universitat Politècnica de Catalunya, Barcelona, Spain
e-mail: fatos@lsi.upc.edu

© Springer International Publishing AG 2017
A. Abraham et al. (eds.), *Computational Intelligence in Wireless Sensor Networks*,
Studies in Computational Intelligence SCI 676, DOI 10.1007/978-3-319-47715-2_2

1   Introduction

Wireless Mesh Networks (WMNs) [1, 2] are gaining a lot of attention because of their low cost nature that makes them attractive for providing wireless Internet connectivity. A WMN is dynamically self-organized and self-configured, with the nodes in the network automatically establishing and maintaining mesh connectivity among themselves (creating, in effect, an ad hoc network). This feature brings many advantages to WMNs such as low up-front cost, easy network maintenance, robustness, and reliable service coverage. Moreover, such infrastructure can be used to deploy community networks, metropolitan area networks, municipal and corporative networks, and to support applications for urban areas, medical, transport and surveillance systems. The main issue of WMNs is to achieve network connectivity and stability as well as Quality of Service (QoS) in terms of user coverage. This problem is very closely related to the family of node placement problems in WMNs, among them, the mesh router mesh nodes placement. Node placement problems have been long investigated in the optimization field due to numerous applications in location science (facility location, logistics, services, etc.) and classification (clustering) [1, 3–6].

WMNs (also referred to as a form of wireless ad hoc network) are in fact one particular type of Wireless Sensor Networks (WSNs) [7], whose topology can vary from a simple star network to an advanced multi-hop wireless mesh network. The main topological feature in this case is that nodes are organized in a mesh topology, making WMNs a reliable infrastructure through redundancy of multi-hop communications.

Different optimization problems can be formulated based on the objectives to optimize and a set of different constraints, such as topological restrictions, battery restrictions, QoS requirements, etc. Some optimization problems are related to minimize the cost of the WMN, such as minimizing the number of mesh router nodes to deploy, while others focus on the WMN performance, such as computing optimal placement of an a priori fixed number of mesh router nodes. The presence of many objectives is in fact a main challenge. These objectives include minimizing the number of mesh routers, maximizing network connectivity, maximizing user coverage, minimizing energy consumption (especially in wireless and mobile networks), minimizing communication delay, maximizing throughput, minimizing deployment cost, etc. And, additionally, there could be certain constraints to take into account such as topological restrictions of the geographical area, interference model, etc. It should also be noted that some of the objectives are contradicting, in the sense that trying to optimize some objective goes in detriment to the optimization of another objective.

Several optimization problems are showing their usefulness to the efficient design of WMNs. These problems are related, among others, to optimizing network connectivity, user coverage and stability. The resolution of these problems turns out to be crucial for optimized network performance. Most important optimization problems in WMNs deal with computing optimal placement of nodes (mesh router nodes, gateways and distribution of mesh client notes), so that network performance is optimized. However, node placement problems are known for their hardness to solve to optimality and therefore heuristics methods are used to nearoptimally solve such problems [8–12].

Given the complexity of node placement problems, most authors have proposed the use of simple heuristic methods or more advanced search methods such as Genetic Algorithms (GAs). So far, only single optimization versions have been considered for the problem. We have considered the bi-objective case, and plan to extend the model to integrate more objectives resulting in a multi-objective optimization model where different objectives could as well be contradicting ones.

In this work, we use HC heuristic method for solving node placement problems in WMNs. We exemplify the applicability of heuristic methods for the case of solving mesh router nodes problem near-optimally.

The rest of the chapter is organized as follows. In Sect. 2 are presented the application scenarios of WMNs. The mesh router nodes placement problem is defined in Sect. 3. In Sect. 4 are described resolution methods for solving nodes placement problem and an introduction of Hill Climbing (HC) algorithm for mesh router placement. In Sect. 5 is described the Web Interface for simulating mesh router nodes placement. The simulation results are given in Sect. 6. In Sect. 7, we give conclusions and future work.

2 Application Scenarios of WMNs

There are a number of application scenarios for which the use of WMNs is a very good alternative to offer connectivity at a low cost.

2.1 Neighboring Community Networks

In a community, the usual solution is to deploy ADSL or cable. However, there are a number of limitations that WMNs can improve as shown in following.

- A large percentage of areas between the houses could not receive wireless services.
- A broadband gateway between different houses could not be shared and wireless services should be established individually.
- A single path to each neighbor can communicate with the rest of neighbors or with the outside.

2.2 Corporative Networks

This scenario corresponds to having a small network for an office or a medium sized network for all offices of a building or even a network to communicate offices located in different buildings. Other similar scenarios include airports, hotels, shopping centers or sports centers.

2.3 Metropolitan Area Networks

Deploying WMNs in metropolitan areas has a number of advantages. The physical layer provides a higher average transmission to any cellular network and need not depend on a wiring. Also, deploying such infrastructure is much cheaper than cable or fiber and can be easily and rapidly deployed in areas with few resources, which have never had any network before.

2.4 Other Scenarios

There are many more scenarios for which WMNs can be used. We mention some of them in following [3].

- *Transportation Systems* Provide information services to passengers, remote monitoring of vehicle safety and communications by the driver.
- *Automatic Control Buildings* In buildings there are several electrical devices to be controlled, including light, elevator, air conditioning, and so on.
- *Medical and Health Systems* In a hospital information monitoring and diagnosis must be transmitted from one room to another.
- *Surveillance* In corporate buildings, shopping malls and stores need broadband data transmission (images and videos basically) for monitoring and surveillance purposes.

3 Mesh Router Nodes Placement Problem in WMNs

Different optimization problems can be formulated based on the objectives to optimize and a set of different constraints, such as topological restrictions, battery restrictions, QoS requirements, etc. Some optimization problems are related to minimize the cost of the WMN, such as minimizing the number of mesh router nodes to deploy, while others focus on the WMN performance, such as computing optimal placement of an a priori fixed number of mesh router nodes [13–15].

The presence of many objectives is in fact a main challenge. These objectives include minimizing the number of mesh routers, maximizing network connectivity, maximizing user coverage, minimizing energy consumption (especially in wireless and mobile networks), minimizing communication delay, maximizing throughput, minimizing deployment cost, etc. And, additionally, there could be certain constraints to take into account such as topological restrictions of the geographical area, interference model, etc. It should also be noted that some of the objectives are contradicting, in the sense that trying to optimize some objective goes in detriment to the optimization of another objective.

In our work, we consider the optimization of mesh router nodes placement in WMNs. In this problem, we are given a grid area arranged in cells where to distribute a number of mesh router nodes and a number of mesh client nodes of fixed positions (of an arbitrary distribution) in the grid area. The objective is to find a location assignment for the mesh routers to the cells of the grid area that maximizes the network connectivity and client coverage. Network connectivity is measured by the size of the giant component of the resulting WMN graph, while the user coverage is simply the number of mesh client nodes that fall within the radio coverage of at least one mesh router node.

An instance of the problem consists as follows.

- N mesh router nodes, each having its own radio coverage, defining thus a vector of routers.
- An area $W \times H$ where to distribute N mesh routers. Positions of mesh routers are not pre-determined, and are to be computed.
- M client mesh nodes located in arbitrary points of the considered area, defining a matrix of clients.

It should be noted that network connectivity and user coverage are among most important metrics in WMNs and directly affect the network performance.

In this work, we have considered a bi-objective optimization in which we first maximize the network connectivity of the WMN (through the maximization of the size of the giant component) and then, the maximization of the number of the user coverage.

For optimization problems having two or more objective functions, two settings are usually considered: the hierarchical and simultaneous optimization. In the former, the objectives are classified (sorted) according to their priority. Thus, for the bi-objective case, one of the objectives, say $f1$, is considered as a primary objective and the other, say $f2$, as secondary one. The meaning is that we first try to optimize $f1$, and then when no further improvements are possible, we try to optimize $f2$ without worsening the best value of $f2$. In the case of WMNs, the hierarchical approach is used due achieving network connectivity is considered more important than user coverage. It should be noted that due to this optimization priority, some client nodes may not be covered due the user coverage is less optimized.

4 Resolution Methods for Solving Nodes Placement Problem

Given the complexity of node placement problems, most authors have proposed the use of simple heuristic methods or more advanced search methods such as Genetic Algorithms [6, 14, 16–21]. We exemplify the applicability of heuristic methods for the case of solving mesh router nodes problem. We have considered a local search method, HC.

4.1 Exact Algorithms

Brute force algorithms These algorithms, also known as enumerative algorithms, can be used to find the optimal solutions (e.g. [14]); however, the solution space of the problem is in general exponentially large and such methods fail to find a solution in reasonable time as they make an exhaustive search of the solution space.

Integer Linear Programming Mathematical programming has been among most used methods in combinatorial optimizations. Its version of integer variables (Integer Linear Programming ILP) has shown useful in modelling and resolution of node placement problems in general and that of node placement in WMN (see e.g. [3, 13]). The main drawback again is that solving an ILP is intractable and can be solved only for small or moderate size instances of the problem.

4.2 Local Search Algorithms

Local Search (LS) algorithms are among the best candidates for solving mesh node placement problems due to their efficiency and simplicity. LS has been used in [16] for the mesh node placement. We present next the application of HC for solving mesh router nodes problem.

Hill Climbing Algorithm for Mesh Router Node Placement We present here the particularization of the Hill Climbing algorithm (see Algorithm 1) for the mesh router node placement problem in WMNs.

Algorithm 1: Hill Climbing algorithm for maximization of f (fitness function).

1: **Start**: Generate an initial solution s_0;
2: $s = s_0$; $s^* = s_0$; $f^* = f(s_0)$;
3: **repeat**
4:   **Movement Selection**: Choose a movement $m = select\_movement(s)$;
5:   **Evaluate & Apply Movement**:
6:   **if** $\delta(s, m) \geq 0$ **then**
7:     $s' = apply(m, s)$;
8:     $s = s'$;
9:   **end if**
10:   **Update Best Solution**:
11:   **if** $f(s') > f(s^*)$ **then**
12:     $f^* = f(s')$;
13:     $s^* = s'$;
14:   **end if**
15:   **Return** s^*, f^*;
16: **until** (stopping condition is met)

Initial solution The algorithms starts by generating an initial solution either random or by *ad hoc* methods [22].

Evaluation of fitness function An important aspect is the determination of an appropriate objective function and its encoding. In our case, the fitness function follows a hierarchical approach in which the main objective is to maximize the size of giant component in WMN.

Neighbor selection and movement types The neighborhood $N(s)$ of a solution s consists of all solutions that are accessible by a local move from s. We have considered three different types of movements. The first, called *Random*, consists in choosing a router at random in the grid area and placing it in a new position at random. The second move, called *Radius*, chooses the router of the largest radio and places it at the center of the most densely populated area of client mesh nodes (see Algorithm 2). Finally, the third move, called *Swap*, consists in swapping two routers: the one of the smallest radio situated in the most densely populated area of client mesh nodes with that of largest radio situated in the least densely populated area of client mesh nodes. The aim is that largest radio routers should serve to more clients by placing them in more dense areas.

We also considered the possibility to combine the above movements in sequences of movements. The idea is to see if the combination of these movements offers some improvement over the best of them alone. We called this type of movement *Combination*:

$$< Rand_1, \ldots, Rand_k; Radius_1, \ldots, Radius_k;$$
$$Swap_1, \ldots, Swap_k >,$$

where k is a user specified parameter.

Acceptability criteria The acceptability criteria for newly generated solution can be done in different ways (simple ascent, steepest ascent, or stochastic). In our case, we have adopted the simple ascent, that is, if s is current solution and m is a movement, the resulting solution s' obtained by applying m to s will be accepted, and hence become current solution, iff the fitness of s' is at least as good as fitness of solution s. In terms of δ function, s' is accepted and becomes current solution if $\delta(s, m) \geq 0$. It should be noted that in this definition we are also accepting solutions that have the same fitness as previous solution. The aim is to give chances to the search to move towards better solutions in solution space. A more strict version would be to accept only solutions that strictly improve the fitness function ($\delta(s, m) > 0$).

Algorithm 2: Radius movement.

1: **Input**: Values H_g and W_g for height and width of a small grid area.
2: **Output**: New configuration of mesh nodes network.
3: Compute the most dense $H_g \times W_g$ area and (x_{dense}, y_{dense}) its central cell point.
4: Compute the position of the router of largest radio coverage ($x_{largest\_cov}$, $y_{largest\_cov}$).
5: Move router at ($x_{largest\_cov}$, $y_{largest\_cov}$) to new position (x_{dense}, y_{dense}).
6: Re-establish mesh nodes network connections.

---

5 Web Interface for Simulating Mesh Router Nodes Placement

The Web application [23] follows a standard Client-Server architecture and is implemented using LAMP (Linux + Apache + MySQL + PHP) technology (see Fig. 1). Remote users (clients) submit their requests by completing first the parameter setting. The parameter values to be provided by the user are classified into three groups, as follows.

Parameters related to the problem instance These include parameter values that determine a problem instance to be solved and consist of number of router nodes, number of mesh client nodes, client mesh distribution, radio coverage interval and size of the deployment area.

Parameters of the resolution method Each method has its own parameters. For instance, Simulated annealing has a starting temperature while Genetic Algorithm has population size, etc.

Execution parameters These parameters are used for stopping condition of the resolution methods and include number of iterations and number of independent runs. The former is provided as a total number of iterations and depending on the method is also divided per phase (e.g., number of iterations in a exploration). The later is used to run the same configuration for the same problem instance and parameter configuration a certain number of times.

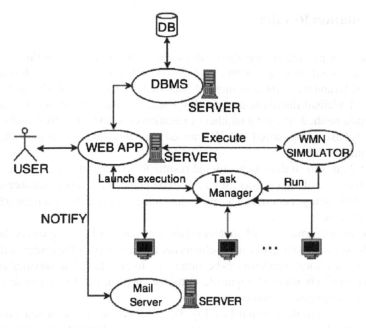

Fig. 1 Web application architecture

5.1 Use Cases

Simulation of different mesh client node positions Client mesh node positions can be chosen from different probability distributions so as to simulate different cases in real life. For instance, positions of stationary client mesh nodes distributed along roadside can be generated using the Exponential distribution while positions of stationary client mesh nodes concentrated at different points in the area can be generated from Weibull distribution.

Simulation of different radio coverage for mesh router nodes Mesh router nodes can have different coverage. The interface allows to select an interval for radio coverage and radio coverage are generated for each router uniformly at random from that interval.

Simulation of different number of mesh routers The number of mesh router nodes is an input to resolution methods. Their positions in the deployment area are to be computed by the resolution methods subject to optimizing WMN metrics.

Simulation of different number of mesh clients The number of mesh client nodes is introduced in input as well. Once this number is provided, the simulator generates the positions of the mesh client nodes using one of the probability distributions (uniform, normal, exponential, Weibull).

6 Simulation Results

The simulation parameters and their values are shown in Table 1. In this work, we considered a small grid area with size (32 × 32). The number of mesh routers is considered 16 and the number of mesh clients 48. We used Normal, Uniform, Exponential and Weibull distribution of mesh clients. For the simulations we used the combination method. The total number of iterations is considered 5000 and the iterations per phase is considered 100. We carried out many simulations to evaluate the performance of WMNs.

In Fig. 2 are shown the computational results for Normal distribution. In Fig. 2a are shown the simulation results for size of giant component versus number of generations. After few generations, HC algorithm achieved to establish a network of all routers connected.

The number of covered mesh routers versus the number of generations is shown in Fig. 2b. In Normal distribution, mesh clients are concentrated at the center of the grid area and the position of mesh routers becomes easy to calculate. The maximal number of covered mesh clients is 43. Figure 2c visualize the position of mesh routers, mesh clients, their connectivity and coverage.

For the Uniform distribution (see Fig. 3), mesh clients are scattered inside the grid area and the positioning of mesh routers becomes difficult. The size of giant component is not maximized and the number of covered mesh clients is 27. In order to cover more mesh clients, the number of mesh routers should be increased.

The simulation results for exponential distribution are shown in Fig. 4. Exponential distribution of mesh clients is similar with a real scenario where stationary mesh clients are distributed along the roadside. For this scenario, all 16 mesh routers are connected with each other and 37 mesh clients are covered.

For Weibull distribution (see Fig. 5) the giant component is maximal and the covered mesh clients are 42.

Table 1 Simulation parameters and their values

Parameters	Values
Clients distribution	Normal, uniform, exponential, Weibull
Number of mesh clients	48
Number of mesh routers	16
Grid size	32 × 32
Router radius	2
Iterations per phase	100
Total number of iterations	5000
Number of generations	50
Apply method	Combination

Fig. 2 Simulation results for normal distribution

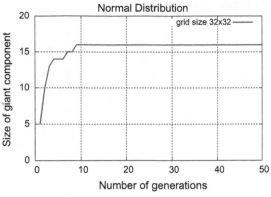

(a) Size of Giant Component

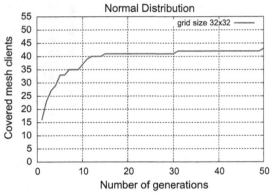

(b) Number of covered mesh clients

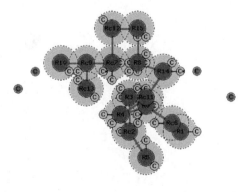

(c) Visualization

Fig. 3 Simulation results for
uniform distribution

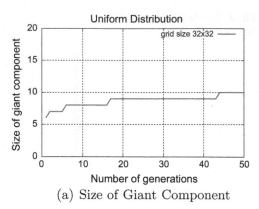

(a) Size of Giant Component

(b) Number of covered mesh clients

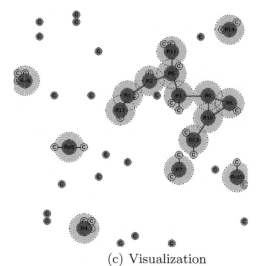

(c) Visualization

Fig. 4 Simulation results for exponential distribution

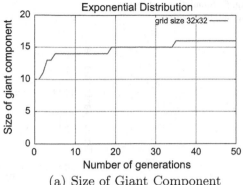

(a) Size of Giant Component

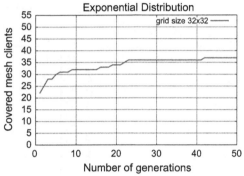

(b) Number of covered mesh clients

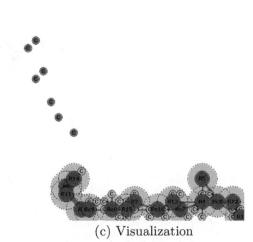

(c) Visualization

Fig. 5 Simulation results for
Weibull distribution

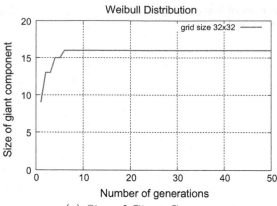

(a) Size of Giant Component

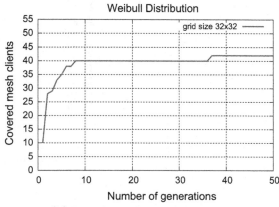

(b) Number of covered mesh clients

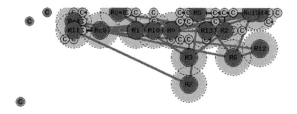

(c) Visualization

From the computational results we can see the performance of HC for different distributions of client mesh nodes. Best performance was achieved for instances with Normal distributions of clients. In particular, we can observe that the uniform distribution causes premature convergence. Finally, it's worth observing that the results corresponding to the Weibull and Exponential distributions are rather similar, which is reasonable since Weibull generalizes the Exponential distribution.

7 Conclusions

This work reveals and stresses the need to consider a plethora of resolution methods, such as exact methods, local search methods and population-based methods, to take advantage of their distinguished features such as exploitation versus exploration of solution space to efficiently solve the optimization problems in WMNs.

In this work we present some optimization problems in WMNs and HC algorithm for solving the problem of mesh router nodes placement in Wireless Mesh Networks (WMNs). In this problem, we are given a number of client mesh nodes a priori distributed in a grid area and a given number of mesh router nodes are to be deployed in the grid area. We formulate the problem as bi-objective optimization problem consisting in the maximization of the size of the giant component in the mesh routers network (for measuring network connectivity) and that of user coverage. We have also presented some experimental results from the HC method using a WMN simulator for small grid area size using different distributions of mesh node clients (Uniform, Normal, Exponential and Weibull). The main challenge identified here is the multi-objective nature of the node placement problems in WMNs. So far, only single optimization versions have been considered for the problem. We have considered the bi-objective case, and plan to extend the model to integrate more objectives resulting in a multi-objective optimization model where different objectives could as well be contradicting ones.

References

1. Akyildiz, I.F., Wang, X., Wang, W.: Wireless mesh networks: a survey. Comput. Netw. **47**(4), 445–487 (2005)
2. Tang, M.: Gateways placement in backbone wireless mesh networks. Int. J. Commun. Netw. Syst. Sci. **1**, 1–89 (2009)
3. Chen, C., Chekuri, C.: Urban wireless mesh network planning: the case of directional antennas. Tech Report No. UIUCDCS-R-2007-2874, Department of Computer Science, University of Illinois at Urbana-Champaign (2007)
4. Zhou, P., Manoj, B.S., Rao, R.: A gateway placement algorithm in wireless mesh networks. In: Proceedings of the 3rd International Conference on Wireless Internet (Austin, Texas, October 22–24, 2007). ICST (Institute for Computer Sciences Social-Informatics and Telecommunications Engineering), ICST, Brussels, Belgium, pp. 1–9 (2007)

5. Vanhatupa, T., Hannikainen, M., Hamalainen, T.D.: Genetic algorithm to optimize node place-ment and configuration for WLAN planning. In: Proceedings of 4th International Symposium on Wireless Communication Systems, pp. 612–616 (2007)
6. Nandiraju, N., Nandiraju, D., Santhanama, L., He, B., Wang, J., Agrawal, D.: Wireless mesh networks: current challenges and future direction of web-in-the-sky. IEEE Wirel. Commun. **14**, 79–89 (2007)
7. Shamshirband, S., et al.: D-FICCA: a density-based fuzzy imperialist competitive clustering algorithm for intrusion detection in wireless sensor networks. Measurement **55**, 212–226 (2014)
8. Amaldi, E., Capone, A., Cesana, M., Filippini, I., Malucelli, F.: Optimization models and methods for planning wireless mesh networks. Comput. Netw. **52**, 2159–2171 (2008)
9. Shojafar, M., et al.: Improving channel assignment in multi-radio wireless mesh networks with learning automata. Wirel. Pers. Commun. **82**(1), 61–80 (2015)
10. Garey, M.R., Johnson, D.S.: Computers and Intractability: A Guide to the Theory of NP-Completeness. Freeman, San Francisco (1979)
11. Lim, A., Rodrigues, B., Wang, F., Xua, Z.: k-Center problems with minimum coverage. Theoret. Comput. Sci. **332**, 1–17 (2005)
12. Wang, J., Xie, B., Cai, K., Agrawal, D.P.: Efficient mesh router placement in wireless mesh networks. In: Proceedings of IEEE MASS07, pp. 1–9 (2007)
13. Li, F., Wang, Y., Li, X.-Y., Nusairat, A., Wu, Y.: Gateway placement for throughput optimization in wireless mesh networks. Mob. Netw. Appl. **13**(1–2), 198–211 (2008)
14. Muthaiah, S.N., Rosenberg, C.: Single gateway placement in wireless mesh networks. In: Proceedings of 8th International IEEE Symposium on Computer Networks, Turkey (2008)
15. Wong, J., Jafari, R., Potkonjak, M.: Gateway placement for latency and energy efficient data aggregation. In: Proceedings IEEE LCN, pp. 490–497 (2004)
16. Antony Franklin, A., Siva Ram Murthy, C.: Node placement algorithm for deployment of two-tier wireless mesh networks. In: Proceedings of IEEE GLOBECOM 2007, IEEE Global Communications Conference, pp. 4823–4827 (2007)
17. Xhafa, F., Snchez, C., Barolli, L.: Genetic algorithms for efficient placement of router nodes in wireless mesh networks. In: AINA, pp. 465–472 (2010)
18. Xhafa, F., Snchez, C., Barolli, L., Miho, R.: An annealing approach to router nodes placement problem in wireless mesh networks. In: CISIS, pp. 245–252 (2010)
19. Barolli, A., Xhafa, F., Takizawa, M.: Optimization problems and resolution methods for node placement in wireless mesh networks. In: 14th International Conference on Network-Based Information Systems (NBiS), pp. 126–134 (2011)
20. Glover, F.: Future paths for integer programming and links to artificial intelligence. Comput. Oper. Res. **5**, 533549 (1986)
21. Holland, J.: Adaptation in Natural and Artifitial Systems. University of Michigan Press, Ann Arbor (1975)
22. Xhafa, F., Sanchez, C., Barolli, L.: Ad hoc and neighborhood search methods for placement of mesh routers in wireless mesh networks. In: Proceedings of ICDCS Workshops of the IEEE 29th International Conference on Distributed Computing Systems (ICDCS09), pp. 400–405 (2009)
23. Xhafa, F., Bravo, A., Barolli, A., Takizawa, M.: An interface for simulating node placement in wireless mesh networks. In: 15th International Conference on Network-Based Information Systems, pp. 326–334 (2012)

An Automated Irrigation System Based on a Low-Cost Microcontroller for Tomato Production in South India

Prabu Mohandas, Arun Kumar Sangaiah, Ajith Abraham
and Jerline Sheebha Anni

Abstract Agriculture plays a major role in India which is required for growth of food crops, intensive farming and rotation of crops. About 70% of people are involved in agriculture. Since, substantial irrigation is necessary for better production in arid regions, reduce the water loss via evapotranspiration is a key to reach sustainable irrigation. This chapter presented a practical result on irrigation controller for the cultivation of vegetable plants (such as tomato) based on fuzzy-logic approaches. The system consists of a feedback fuzzy logic controller that records key parameters with sensor, Zigbee–GPRS remote monitor and a database. The system is simple to install in existing micro irrigation systems without modifying the existing. Based on the crop yield, the fuzzy logic controller acquires data from the sensors and applies fuzzy rules to acquire suitable time for irrigation. All variables are fuzzified using trapezoidal and triangular membership functions in MATLAB. In this fuzzification, a Max–Min inference engine and a Mamdani-type rule base was adopted in order to make the best decision for each situation. Thus by preventing needless irrigation, not only water demand is reduced, but it is also possible to ensure the protection of freshwater resources. The system is developed and tested with growth of vegetable plants (tomato). It saves 50–60% water utilization as well as the energy generation cost. A local farmer (through mobile) saves real-time data received from the field controller via wireless Zigbee protocol and transmit the collected data to a remote station via a GPRS link. This enhancement enables tracking analysis and improvement of system performance in real time. The deployment of fuzzy control combined with remote data logging would foster better management of irrigation in arid lands

P. Mohandas
Department of Computer Science and Engineering,
Adhiyamaan College of Engineering, Hosur, India

A.K. Sangaiah (✉)
School of Computing Science and Engineering, VIT University, Vellore, India
e-mail: arunkumarsangaiah@gmail.com

A. Abraham
Machine Intelligence Research Labs (MIR Labs), Auburn, WA, USA

J.S. Anni
BTI College of Engineering, Bangalore, India

© Springer International Publishing AG 2017   49
A. Abraham et al. (eds.), *Computational Intelligence in Wireless Sensor Networks*,
Studies in Computational Intelligence SCI 676, DOI 10.1007/978-3-319-47715-2_3

such as hill stations of Hosur. Through this method, it is possible to reduce the power, water demand, the total power, the battery and power control unit costs.

Keywords Evapotranspiration · Fuzzy logic controller · Irrigation controller · GPRS remote monitoring · Remote data logging

1 Introduction

Irrigation systems were used by the Egyptians and it is in the history of records. The idea of automated irrigation system is an older technique, human kind has identified to supply water among large areas of foliage through the usage of automated and drip irrigation systems. Automated irrigation systems are efficient to irrigate plants to an appropriate level for a normal growth of plants. These systems can reduce the wastage of water. The irrigation controller is used to monitor the entire irrigation system. It observes the supply of water and provides fertilizer to plants. Therefore the farmer can be able to achieve the optimum quantity of water and fertilizer during the growth. The green house irrigation controller controls the supply of water through computerized controller. Most of the conventional methods are not preferred since it is based on the on-off control methods and proportional control methods. This method results in the reduction of productivity and energy. This chapter presents a practical result on irrigation controller depending on the fuzzy-logic concept. It can be given in two ways. First, it identifies the overall problem of irrigation and it recognizes the physical control model. The Fuzzy Logic Controller (FLC) is based on the Mamdani controller and is developed using MATLAB. The formal presentation of the fuzzy logic controller provides simple construction of system and advantages of using fuzzy in the feedback control.

The fuzzy logic controller which is developed can determine the quantity of water required by plants in well defined depth using evapotranspiration functions, green house environmental conditions, type of soil, plant type and other factors affecting the green house irrigation. Many technologies are developed in the agriculture field such as automatic crop monitoring systems, automatic irrigation scheduling systems and automatic data-acquisition system. Automatic crop monitoring system reaches temperature and moisture level more quickly than the manual system. Automatic irrigation scheduling system can be used to supply the water automatically based on the measurements specified through sensors. Automatic data-acquisition systems can be used to collect information such as condition of soil moisture, soil temperature, irrigation duration and air temperature. There are many advantages in these systems. Sensors fixed in the crop fields are used to ensure the water level in the soil. A solution to the problem of excessive supply of water can prevent damage of crops. Data can be collected continuously during the cropping season. Manual irrigation does not provide these benefits.

In this section, we review the earlier work on the various irrigation systems implemented in different applications using diverse algorithms. Javadi et al. [10] have proposed a system for irrigation based on control system of fuzzy methodology. The FLC model estimates effectively the quantity of water required by plants in well defined depth using good irrigation model evapotranspiration functions, green house environmental conditions, type of soil, plant type and other factors which affects the green house irrigation. Sabrine et al. [16] have presented a thorough irrigation result to the farmers depending on WSN. An automated irrigation system using sensor nodes of lower cost with lower power consumption can reduce water waste and is cost effective. A node is deployed using telosb mote and sufficient sensors. Field nodes can be used to determine the moisture and temperature level in the soil. Weather nodes can be used to monitor climatic changes, and nodes connected to actuators can be used to control the valve when needed.

The literature Survey on water irrigation is discussed in the following section and shown in Table 1.

Table 1 Earlier studies in the context of proposed research

Author's name Name of the article and year of publication	Methodology	Attainment
Genghuang et al. [8] Automation irrigation system based on wireless network, 2010	GSM (Global System Mobile) network and radio communication	Orders can be sent to the controller and the information are sampled by the controller can also be sent to the cell phone by GSM message
Guang et al. [9] Growth and Comprehensive Quality index of tomato under Rain Shelters in Response to Different Irrigation and Drainage Treatments, 2014	Underground sensors	Drainage treatments increased the average yield
Javadi et al. [10] Intelligent Control based Fuzzy logic for Automation of Greenhouse Irrigation system and evaluation in Relational to conventional systems, 2009	Fuzzy Logic Controller (FLC) prototype is based Mamdani controller using MATLAB	It can save a lot of water, very cheap to Implement the water irrigation system
Joaquin et al. [11] Automated Irrigation System using a wireless sensor network and GPRS Module, 2014	Photovoltaic panels and duplex communication link based on cellular-Internet	Feasible and cost effective for optimizing water resources, used for agricultural production
Robert et al. [15] Wireless sensor Network with irrigation valve control Computers and Electronics in Agriculture, 2013	Valve Actuation System: Time based and schedule based irrigation systems	A web based interface was used for manual control of irrigation valves and displayed the status of all actuators in the network
Sabrine et al. [16] Precision irrigation based on wireless sensor network, 2014	low-cost sensor nodes having reduced power consumption able to realize necessary requirement	Recorded in base station and sent to the farmer's PC just in time to allow him to take the proper action

The earlier studies in the context of proposed research are presented in Table 1 as illustrated in the following section.

Robert et al. [15] promoted a commercial wireless sensor and control network using hardware and software to control valve. The valve positioning sensor system includes custom node firmware, sensor hardware and firmware, control for Internet gateway and communication software for web interface. The system uses radio range of single hop using a mesh network with 34 valve sensors, controlling the valves and water meters. Christos et al. [2] explained the design of an adaptable support system and its integration with wireless sensor network to implement autonomous closed-loop zone-specific irrigation. Ontology can be used to define application logic system's flexibility, adaptability and its supports the application of automatic inferential and validation mechanisms.

Evans [6] focused on the data acquisition and control mechanism for automatic irrigation of central pivot and linear move systems. The implementation of Power Line Carrier (PLC) provides low cost communication among distributed control systems. The various sensor types (tensiometers and GMS (Granular Matrix Sensors), switching tensiometers, dielectric probe which uses a voltage signal) to obtain automatic irrigation control was discussed by Rafael et al. [14]. They found that a 70 % reduction in water use in automatic irrigation when compared to manual irrigation practices.

Daniel et al. [3] developed low cost micro controller based system to maintain the crop condition by examining the parameters such as canopy temperature, soil temperature, soil moisture and air temperature. This system is used to provide automated measurement with valid performance. The user can be able to download the data from the Internet. Shaughnessy et al. [17] proposed a TTT (Time Temperature Threshold) algorithm for auto irrigation scheduling. They have used this algorithm to an automated center point irrigation system and they analyzed it both manually and automatically. Manually scheduled irrigation was based on neutron probe readings. They have also analyzed the crop yields and water use efficiency for manually and automatically irrigated cotton plots in 2007 and 2008.

Kati et al. [12] developed an experimental method for irrigation scheduling system. The volume of water, measurements of plants and fruits are measured in South Florida. Set schedule irrigation system requires more amount of water when compared to the other systems such as evapotranspiration and soil water sensor based irrigation system. The amount of water vapor, transpiration, efficient usage of water and evapotranspiration has been reviewed. The fruit measurements are also reviewed through plant fruit number and weight per harvest in response to irrigation treatment. Finally they have identified 65 % of water can be saved through evapotranspiration or soil water sensor based sensor scheduling.

Genghuang et al. [8] and Guang et al. [9] illustrated automated irrigation system has three levels such as pc control platform, controller and action unit. Irrigation plan to the controller can be transmitted through mobile phone and pc controller platform.

Using mobile phone we can access the information such as time, temperature, soil moisture, air humidity. To open or close the valve of nozzle the controller sends the control command to action unit based on the irrigation plan.

Xiaohong et al. [18] described a system which contains sensor node, coordinator node and irrigation controller node. Sensor node is fixed in the crop field to collect data about temperature of soil and humidity. The collected data will be transmitted to the coordinator node which contains fuzzy controller. Temperature of soil and humidity is given as the input to the fuzzy controller. The output will be given to an irrigation controller node which supplies water to the crop field. By using fuzzy logic controller the system saves water for irrigation purpose.

Model Predictive Control (MPC) technique to compute optimal control strategies for actuators was proposed by Yang et al. [20]. The actuators were controlled by a remote station according to the measurements of sensors. The proposed technique optimizes the system performance by reducing cost, maintaining water storage levels in dams and reservoirs, quality management etc. Joaquin et al. [11] proposed an automated irrigation system using low cost microelectronic components. The automation of watering depends on the measurement of soil moisture and temperature. This system consists of a GPRS module that uploads the information about soil moisture, temperature and water supplied.

A system is needed for ongoing monitoring to prevent crop damage and to improve crop yield. The objective is to construct an inexpensive microcontroller system for measuring soil moisture and temperature in crop fields. A system is built and tested in plotted plants to evaluate the performance. This design standard includes automated measurements of substrate water content in the plotted plants and the measurements are automated to the web page through GPRS. Irrigation systems such as sprinkler, drip, center-pivot etc. Can be easily automated if, these systems incorporate new sensor technologies and latest microelectronic components.

Section 1 describes the automated irrigation system discussed by various authors, fuzzy logic Controller based on irrigation system were discussed. Section 2 discusses the sensing and control unit and Fuzzy logic control unit.

Section 3 demonstrates the experimental setup. Section 4 explains the result and discussion. Conclusion and future enhancements of this work were discussed in Sect. 3.

2   System Design

The system structure of Automatic Irrigation System consists of two main systems,

 (i) Sensing and control
(ii) Fuzzy logic control

2.1 Sensing and Control

The controller consists of specific field sensors, an FLC-embedded data logging and wireless communication board, a flow sensor and power supply system.

2.1.1 Field Sensors and Calibrations

Three sensors are used for monitoring the environment, the microcontroller, soil moisture (VG400RevD), and Temperature sensor (DS18B20). These sensors are considered most critical in the irrigation process and their readings are used by the FLC to devise the different fuzzy rules. The other available parameters like relative humidity in arid lands have little effect.

2.1.2 Soil Moisture (VG400RevD) and Temperature Sensor (DS18B20)

The VG400RevD sensor delivers a maximum of 3 V for 100 % moisture when powered with 5 V. The soil moisture sensor is inserted in the root zone of the plotted plants to estimate the soil moisture. The threshold value is set in the soil moisture, when the soil humidity exceeds D0 (digital output interface) output is low. When the soil humidity drops below threshold value, D0 output is high [7].

In order to get the precise value of soil moisture, an analog output A0 and AD (analog-to-digital) module is connected through an AD converter. The sensor is monitored by a microcontroller through an ADC port and powered at 3 V. The digital thermometer, model DS18B20 (maxim integrated) is used for obtaining air temperature measurements. The DS18B20 supports 9-bit to 12-bit Celsius temperature measurements, 1-wire bus protocol. The thermometer has $\pm 0.5\,^\circ$C accuracy on a range from -10 to $+8\,^\circ$C temperature range and a unique 64-bit serial code. The sensor converts the temperature to 12-bit digital word in 750 ms.

2.1.3 Micro Controller (LPC 2148)

LPC2148, 32-bit microcontroller with 64-pins operates in a range 3.0–3.6 V at 30MHZ with internal oscillator. It has 45 digital input/output ports, 10-bit analog-to-digital converters, multiple serial interfaces which include two UARTS (16C550), two fast I2C-bus (400 kbit/s), two 32-bit timers, low power Real-Time Clock (RTC), 512KB flash memory, 32KB of static RAM. It supports both the idle mode and power-down mode.

2.1.4 XBee Radio Transceiver (XB24-Z7WIT-004)

The ZigBee device used in this system is XBee (Digi International Inc. [4], Xin [19]). It is based on IEEE802.15.4 standard (Wireless Personal Network). It supports low-power Wireless Sensor Networks and operate up to a distance of 120 m line-of-sight with 250 kbps RF (Radio Frequency) data rate. It is powered at 2–6 V and interfaced to the microcontroller through serial port. Xbee radio device operates at 2.4 GHz frequency.

2.1.5 Microcontroller/Microprocessor Interface (LCD HD44780)

The microcontroller/microprocessor interface to HD44780 LCD modules (hereafter generically referred to as character LCD modules) is almost always 14 pins. Some displays have additional pins for backlighting or other purposes, but the first 14 pins still serve as the interface. The first three pins provide power to the LCD module. Pin 1 is GND and should be grounded to the power supply. Pin 2 is VCC and should be connected to +5 V power. Pin 3 is the LCD Display bias. By adjusting the voltage or duty cycle of pin 3, the contrast of the display can be adjusted. Most character LCDs can achieve good display contrast with a voltage between 0 and 5 V on pin 3.

2.2 FLC Based Data Logging and Wireless Controller

The board is constructed around the microcontroller (LPC 2148) which makes the system compact, deployable, adaptive, and scalable. The main tasks of this board are data acquiring, processing logging, and transmitting. An application-specific FLC is devised and loaded to the microcontroller. MAT LAB simulations and system reduction are performed on the FLC beforehand.

Fuzzy sets and logic can handle real-life uncertainties, hence it is ideal for such nonlinear, time-varying and hysteretic control system in Bellazzi et al. [1], Do Guen et al. [5]. A detailed description of the FLC is given in the subsection below. Also, the board is enhanced with Zigbee modem for short-range wireless connectivity. In this application, the board is configured to transmit all sensors' reading to the local station every 30 min. The local station saves and in turn transmits processed data to the remote station. This feature enables remote monitoring and creation of database for irrigation parameters, environmental conditions, and system performance for analysis and better management. In this application to schedule irrigation, the FLC uses threshold moisture of 30 % (see Fig. 1), which is within the sensor range.

Figure 1 explains the structure of the FLC implemented here. All variables are fuzzified using trapezoidal and triangular membership functions. The membership

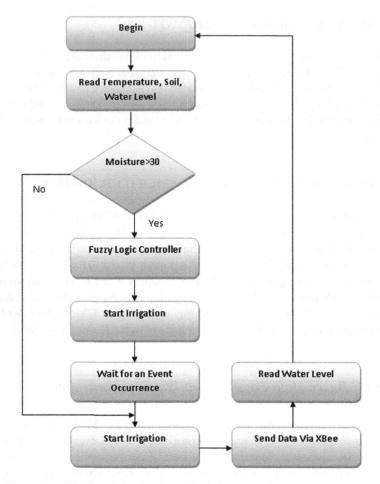

Fig. 1 System flow: FLC based data logging and wireless controller

functions are distributed according to the possible value of each variable after fuzzification. The Fuzzification process, Max–Min inference engine and Mamdani-type
rule base produce the required decision for each situation. After application of a
centroid defuzzification, the controller produces the desired output.

2.3 Fuzzy Experts and Interference System

Human reasoning can grasp uncertain concepts appropriately. However, it cannot be
expressed in precise terms. Fuzzy logic provides a methodology to model uncertain
and human way of thinking reasoning and perception. In Boolean logic, there are
two concepts 'true' and 'false', which are represented by 1 and 0 respectively. This

means any proposition can be true or false. Fuzzy logic is an extension of Boolean logic that allows intermediate values between these two extremes.

Fuzzy systems provide the means for presenting the expert knowledge of humans about the process in terms of fuzzy rules (IF THEN). A fuzzy rule is the basic unit for capturing knowledge in fuzzy systems. Fuzzy inference is the process of mapping from a given input to an output using fuzzy logic. The goal is to obtain a conclusion consisting of one or more consequents from a premise consisting of one or more antecedents. By mapping, decisions can be made or patterns recognized. The process of fuzzy inference involves membership functions, fuzzy logic operators, and if then rules. There are two types of fuzzy inference systems that can be implemented in the fuzzy logic toolbox: Mamdani-type and sugenotype. These two types vary in their output. A fuzzy rule, like a conventional rule has two components: an 'if part and a 'then' part which are referred to as antecedent and consequent, respectively. The main structure of fuzzy rule is given in Eq. 1

$$IF<antecedent>THEN<consequent> \tag{1}$$

The predecessor of fuzzy logic has condition that should be satisfied by a degree. Typically, the antecedent of a fuzzy logic can merge multiple single condition into complex condition using logic gates such as AND, OR and NOT logical operators. The resultants of fuzzy logic are classified into two major categories: Fuzzy consequent (Eq. 3, in which C is a fuzzy set), functional consequent (Eq. 3, in which p, q and r are constants). Fuzzy Interference System (FIS) integrate an expert's affair to form a system design. They can be comprised of four blocks. A FIS consists of a fuzzifier which will transform the 'crisp' inputs into fuzzy inputs through membership functions which constitute fuzzy sets of input vectors, a knowledge-base which encompass the related information which is given by the expert in the form of linguistic fuzzy rules, an interference system (Engine) which uses them together with knowledge-base for inference by a method of reason and a defuzzifier that modifies the fuzzy results of the interference into a crisp output through defuzzification method.

The knowledge-base consists of two components: a data-base, which defines fuzzy set membership functions which is used in the fuzzy rules, and a rule-base encompassing a collection of linguistic rules that can be joined by some specific operator. Depending on the resulting method of fuzzy rules, there are two common types of FIS, which may change based on the differences between the specifications of the resulting part (Eqs. 2 and 3). The first fuzzy system uses the inference method proposed by Mamdani in which the rule consequence is defined by fuzzy sets and has the following structure [13].

$$IF\ x\ is\ A\ and\ y\ is\ THEN\ f\ is\ C \tag{2}$$

The second fuzzy system proposed by Takagi, Sugeno and Kang (TSK) has an inference engine in which the conclusion of a fuzzy rule comprises a weighted linear combination of the crisp inputs rather than a fuzzy set. The TSK system has the following structure.

$$Fx \ is \ A \ and \ y \ is \ B \ THEN \ px + qy + r \qquad (3)$$

Here p, q and r are constant parameters. TSK models are suitable for approximating a large class of non-linear systems. The knowledge-base containing the database and rule-base of an FIS can be constructed from an expert's knowledge. Fuzzy Interference System (FIS) is the main component of fuzzy logic controller, which consists of five processing parts:

- Linguistic variables can be generated depending on crisp information inputs from sensor subsystem through Fuzzification interface
- Crisp control outputs to the actuators can be generated by defuzzification interface
- Interface operations can be generated based on predefined control logic through decision making unit
- Fuzzy sets and membership functions which is used in fuzzy rules can be provided by database process
- Sufficient number of fuzzy rules can be comprised by rule base unit.

Based on considered membership functions for inputs, the Mamdani fuzzy rule based system has $3 \times 3 = 9$ rules. The system is implemented using the following FIS properties:

Type: 'Mamdani'
Decision method for fuzzy logic operators AND (intersection): 'MIN'
Decision method for fuzzy logic operators OR (union): 'MAX'
Implication method: 'MIN'
Aggregation method: 'MAX'
Defuzzification: 'CENTROID' (center of gravity)

2.3.1 Fuzzy Rules Determination

Many investigators have examined the technique for formatting the rules and skilled person's knowledge is the one most commonly used. The expert is asked to recap the familiarity about the system in the form of a basis and outcomes. The rules and rule determinations are based on fuzzy classifier techniques. The irrigation system of fuzzy model was deliberated by a set of rules based on the expert knowledge. For the input, the Member Function Interference System (MFIS) used here has $3 \times 3 = 9$ rules based on the membership functions. An example of rule description is if Soil Moisture is "Cold" then Water Flow is "Low". The description of the rules is shown in Tables 2, 3 and 4.

Table 2 Rules for duration of irrigation

Soil moisture Condition/status	Temperature sensor (C)		
	Dry	Medium	Wet
Cold			✓
Medium		✓	
Hot	✓		

Table 3 Rules for duration of irrigation

Soil moisture Condition/status	Water flow		
	High	Medium	Low
Cold			✓
Medium		✓	
Hot	✓		

Table 4 Rules for duration of irrigation

Temperature sensor Condition/status	Water flow		
	High	Medium	Low
Dry	✓		
Medium		✓	
Wet			✓

3 Experimental Setup

The experiment was carried out in Hosur during January 2015. The experimental field was previously plated with tomato crop. The experimental area was extended over 1.5 ha. The experimental area was irrigated and left to attain moisture level ranges of 10–15 db. Different field practices were applied in this research. End to end transmission of data was censured by a Zigbee–GPRS platform. The communication between the field controller and local station was done over a short range (100 m line of sight). IEEE 802.15.4 compliant zigbee protocol (Xbee Series 1 Chip) by virtue of its-low power feature. (The rest of the activity is described in the present tense).

In addition, IEEE 802.15.4 compliant chips present unique features for conventional and precision agriculture. The local stations receives, process and saves data from f ield and then routes this data to a remote station (i.e. PC) via wise coverage GPRS modules on both sides.

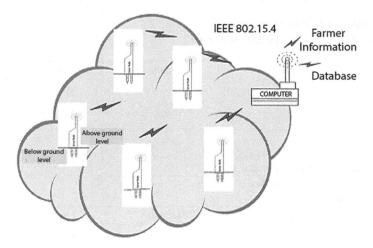

Fig. 2 Irrigation system deployed in Wireless Sensor Networks

The GPRS module interfaces to the remote station via microcontroller over UART. All data are saved and displayed in real-time at the remote station. This experiment consists of two types of sensors such as (1) temperature sensor and (2) moisture sensor. The main objective of this research is to find the water flow to the plant using Moisture sensor. The water flow of the soil is purely equivalent to the moisture. The three main steps are included in the experiment such as follows:

- If the temperature is high, it defines the moisture level of the soil is low i.e., the soil is dry and the plant needs water.
- If the temperature is medium, it is indicates the moisture of the soil is medium, and the plants need no water.
- If the temperature is low, it indicates the soil is wet and there is no need for water until it reaches the dry state.

Here the threshold for moisture of the soil is considered as 30. The moisture above 30 indicates dry condition of the soil. The moisture is medium when the value is between 25–30, and wet when the value is less than 20. The sensor is placed at a distance of 9 cm from the plant. The message/notification from the sensor is passed to the IEEE 802.15.4 which uses to transfer the message to the mobile and also saves the moisture value and time in the database. The overall irrigation systems with deployed roles are shown in Fig. 2.

3.1 Components and Usages

An automated irrigation system acts according to the information read from soil moisture sensors and temperature sensors, which was controlled by the microcontroller. Sensors, XBee devices and external components are connected as a circuit and controlled by the programmable microcontroller. Based on the application, the components were selected. The circuit affords reliable performance, if the components provide desired functions, incorporated easily with the microcontroller, power supply for the entire circuitry, inexpensive and low power microelectronic components. An efficient automated irrigations system was constructed based on two important components. The components are (1) server unit and (2) sensor unit. Both the units are connected by the radio transceivers (XBee), which is used to transfer the recording information from soil moisture sensor and temperature sensor. The GPRS module is used to forward the real time information to a webpage through mobile network located on server unit. This enables remote monitoring of the current status of field's moisture level and crop fields temperature through Internet.

3.2 Design and Operation of Sensor Unit

The sensor units consist of microcontroller, temperature sensor, soil moisture, sensor LCD, XBee, water motor and SPDT relay. The circuit design is based on some other design but modified to have more powerful microcontroller. A photograph of the completed circuit board of sensor node and server node were shown in Figs. 3 and 4. For the real time resistant, the sensors are designed to transmit the current date and time via XBee radio modem through two digital transmitters (TR) and receiver (RX). The list of circuits and components are shown in Table 5.

The electromagnetic switch is used to control the water pump pour level for automatic irrigation, and the irrigation systems carried through SPDT (Single Pole Double Throw) relay operates at 5 V. The Single Pole Double throw enumerates the coil, a common terminal, a closed terminal and an open terminal. The operations are based on the coil aspects.

Fig. 3 Circuit board and sensor node component

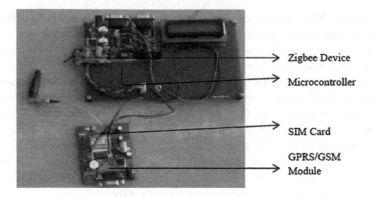

Fig. 4 Circuit board and server node components

Table 5 List of circuit components and sensors

Description	Model number
Microcontroller	LPC2148
Soil moisture sensor	FC-28
Temperature sensor	DS18B20
XBee radio modem	XB24-Z7WIT-004
SPDT relay	JZC-11F
9V DC water pump	EWP-7L9
SIM900 GPRS/modem	TW161
Miscellaneous (oscillator, resistors, capacitors, sockets, cables, batteries, etc.)	

The operations are

- If the coil is not energized then the common terminal and the closed terminal have connection.
- If the coil is energized, the common terminal and the open terminal have connection.

A water pump pour level was controlled by SPDT relay, which was connected to a microcontroller. Two different irrigation events are implemented in the sensor unit. The events are,

- When the soil moisture sensor fall below the threshold level.
- When the temperature sensor go above the threshold level.

For displaying the current status of irrigation (soil moisture level) and the temperature values, LCD is used. An Alphanumeric LCD controller (HD44780) is used with the specification of 20×4 display, 16 pins in all, made up of power control pins, display control pins, data pins, LED+ anode, and LED− cathode.

3.3   Design and Operation of Server Unit

The server node enumerates of an XBee radio modem, LCD, memory (24FC1025) GPRS module (SIM900) and a microcontroller (LPC2148). The distance between the server node and the sensor node is 120 m line-of-sight and the server node is located away from the plotted plants. The server node figure is shown in Fig. 6. The server node microcontroller receives the information regarding the moisture level and current field temperature along with date and time and is stored in the memory (24FC1025). The recorded information is transferred to a web server through GPRS Module.

The SIM900 module is used to transfer the data from field report to the web page. This module will operate at the range of 3.2–4.8 V. The roles of downlink and uplink ranges are 85.6 and 42.8 kbps respectively. It consists of a serial port with control lines and status of modem interface. It creates the connection to URL of the web server to upload and download the recorded data.

Characterizing the Input in Fuzzy Set

For the declaration of inputs, triangular membership function types are used. There are three membership functions used in Temperature variable. The membership functions are "Cold", "Medium", and "Hot", which is plotted in MATLAB fuzzy logic tool box and shown in Fig. 5.

There are three membership functions used in Soil Moisture variable. The membership functions are "Dry", "Medium", and "Wet", which is plotted in MATLAB fuzzy logic tool box and is shown in Fig. 6.

"High", "Medium" and "Low" are used to show the various ranges of input fuzzy variable "Water Irrigation" in a plot consisting of three regions as shown in Fig. 7.

Number of rules = m n

Where   m = maximum number of overlapped fuzzy sets

n = number of inputs

Total number of rules is equal = product of number of functions.

Fig. 5 Temperature membership function

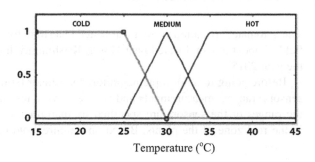

Fig. 6 Soil moisture
membership function

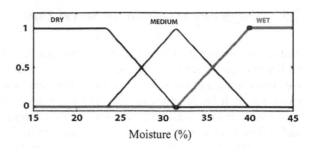

Fig. 7 Water (irrigation)
flow membership function

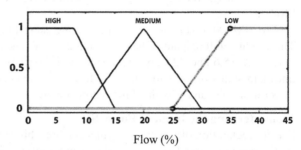

For this fuzzy based irrigation design, m = 3 and n = 3, so the total number of rules are 09.

The two input variables described here consist of three membership functions. Thus, 3 × 3 = 09 rules were required as shown in Tables 2, 3 and 4.

Figure 8 shows the output as (simulated results) very near to estimated value. The results have been inveterate for all possible simulated value of input variables i.e. Temperature level and soil moisture level.

The irrigation system is organized according to soil moisture and Temperature Sensors are obtained by the Fuzzy logic is controlled according to the data coming from the sensors. The surface view distribution is obtained using this kind of irrigation can be seen in Fig. 9.

4 Results and Discussion

The automated irrigation system was tested in a small tomato greenhouse production field of about 500 m, located near Hosur, Krishnagiri district, Tamil Nadu, India in the year 2015.

Before going to field implementation, the sensor (soil moisture and temperature sensors) ranges, measurements and accuracy were tested and examined. Normally, the ranges of the sensors are specified by manufacturers. The sensors are placed in the root zone of the plants. Based on the threshold range, the irrigation system

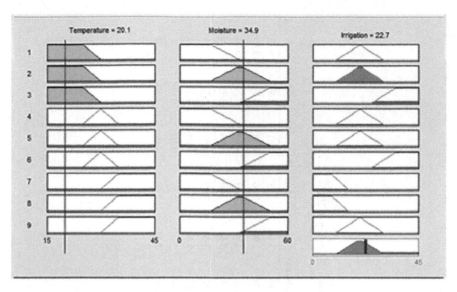

Fig. 8 Fuzzy rule viewer

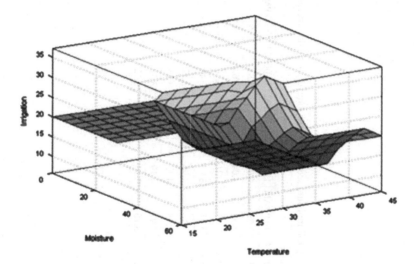

Fig. 9 Surface viewer for the irrigation system

was executed. The horizontal lines are shows that the threshold levels for both the temperature and soil moisture.

The threshold ranges are,

Volumetric Water Content for the soil = 2.5 % (minimum value)

Temperature = 30 °C as threshold level.

Table 6 Statistics of data sampled by soil moisture sensors (%) for tomato production (period: Jan 3, 2015–Jan 6, 2015)

Date	Sensor units	Timings												
		8.00 AM	10.00 AM	12.00 AM	14.00 PM	16.00 PM	18.00 PM	20.00 PM	22.00 PM	24.00 PM	2.00 AM	4.00 AM	6.00 AM
Jan 03, 2015	Unit 1	4.03	3.79	4.01	3.99	3.81	3.66	3.49	3.14	2.98	2.61	2.3	4.13
	Unit 2	3.9	3.52	4.04	3.96	3.68	3.42	3.29	2.99	2.76	2.53	4.1	3.89
Jan 04, 2015	Unit 1	3.7	3.4	4.21	3.98	3.76	3.42	3.13	2.96	2.89	2.64	2.51	3.96
	Unit 2	3.66	3.41	3.91	3.79	3.42	3.29	2.98	2.62	2.46	3.96	3.75	3.49
Jan 05, 2015	Unit 1	3.74	3.58	4.08	3.71	3.52	3.31	2.91	2.67	2.49	3.91	3.72	3.53
	Unit 2	3.27	2.98	4.13	3.9	3.45	3.26	2.97	2.62	2.49	4.04	3.92	3.73
Jan 06, 2015	Unit 1	3.21	2.91	4.19	3.99	3.51	3.19	2.86	2.67	2.51	4.09	3.99	3.76
	Unit 2	3.52	3.16	3.99	3.78	3.52	3.18	2.94	2.73	2.53	4.0	3.89	3.68

Table 7 Statistics of data sampled by temperature sensors (°C) for tomato production (period: Jan 3, 2015–Jan 6, 2015)

Date	Sensor units	Timings											
		8.00 AM	10.00 AM	12.00 AM	14.00 PM	16.00 PM	18.00 PM	20.00 PM	22.00 PM	24.00 PM	2.00 AM	4.00 AM	6.00 AM
Jan 03, 2015	Unit I	24	27	31	29	26	25	24	23	22	23	22	24
	Unit 2	25	29	31	29	25	24	23	22	22	24	23	25
Jan 04, 2015	Unit I	26	28	30	29	27	26	25	23	24	21	24	22
	Unit 2	28	29	31	29	27	24	21	22	21	25	22	26
Jan 05, 2015	Unit I	27	29	31	29	23	21	24	22	23	21	21	25
	Unit 2	29	28	31	29	22	25	23	24	22	21	22	25
Jan 06, 2015	Unit I	27	29	31	28	29	27	25	26	24	22	24	27
	Unit 2	26	28	30	29	28	27	24	25	22	24	23	27

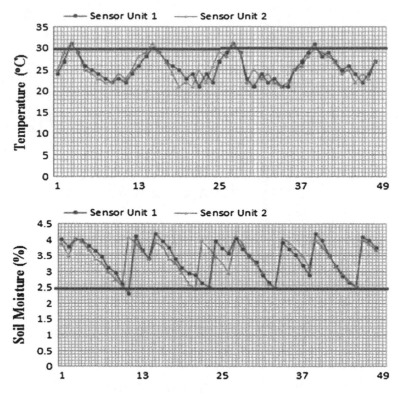

Fig. 10 Gathered data of sensor unit 1 and sensor unit 2 of the automated irrigation system: temperature and soil moisture

(Note: X-axis: 1–13 → Jan 3, 2015; 13–25 → Jan 4, 2015; 25–37 → Jan 5, 2015; 37 to 49 → Jan 6, 2015)

The server unit was placed about 120 m line of sight from the sensor units. These estimated/recorded results were uploaded automatically to the webpage (website) at intervals of two hours for remote monitoring. For occurrence, four days recorded temperature and moisture information of two sensor units are shown in the Tables 6 and 7 respectively.

Four days field soil moisture level and temperature levels are monitored and represented in graphical manner, which are shown in Fig. 10. In Fig. 10, the horizontal bar represents the threshold level of temperature and soil moisture of sensor units. The recorded temperature and soil moisture information from sensor unit 1 is represented as a graph and shown in Fig. 11. The recorded temperature and soil moisture information from sensor unit 2 is represented as a graph and shown in Fig. 12. The vertical bars are represented as threshold level, which was designed and assigned based on current temperature and soil moisture on field.

Typical real-time data collected remotely at two different locations are illustrated in Figs. 10, 11 and 12 for examining the system's behavior. The reliability of these

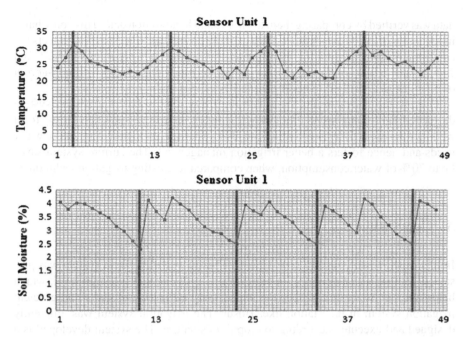

Fig. 11 Automated irrigation (*vertical bars*) activated by the temperature threshold ≤30 °C and soil moisture threshold ≤2.5 % in sensor unit 1

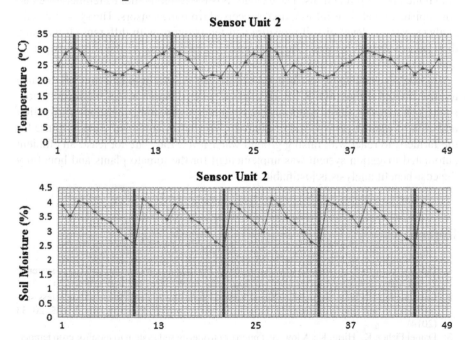

Fig. 12 Automated irrigation (*vertical bars*) activated by the temperature threshold ≤30 °C and soil moisture threshold ≤2.5 % in sensor unit 2

data was verified by comparing them to on-site and local station data. This contributes in reduction of evapotranspiration losses for the entire day. Also, the FLC selects the appropriate time and duration by referring to the base of fuzzy rules (Tables 6 and 7 shows the actual soil moisture tracking the desired value (30 %) with non-abrupt oscillations as compared to traditional feedback systems).

Overall, the above results show that the fuzzy control-based irrigation system developed here compensate efficient water losses in arid regions like Hosur for a well- systematic activity to grow the crop. The system enables predicting future water needs and hence fosters a better irrigation management. The current system saves up to 70 % of water consumption, when compared to existing irrigation techniques.

5   Conclusion

In this paper, an automated fuzzy controller based water irrigation system was presented with regular monitoring of temperature level and soil moisture level, which helps water conservation, reduce the unnecessary utilization of water and efficient irrigation system in hill station like, Hosur. The irrigation system was efficiently designed and executed according to crop field's nature. The system developed is a fuzzy logic controller based on Mamdani fuzzification using triangular and trapezoidal membership functions. The execution is based on the field's temperature and soil moisture level. This information is received from the sensors. The system briefly analyzes various level of soil moisture and temperature with different climate and dissimilar water pour level. Thus, the system enables predict the future irrigation needs.

In order to evaluate the performance of irrigation system in real time all the data are transmitted through GPRS-Zigbee based wireless system. All the information, such as climate condition, soil moisture level along with time, duration of irrigation and utilized water level is stored in database. This information will help in future for the farmers, to react with raising ground water level. Last but not least, an efficient automated irrigation system was implemented for the tomato plants and hopefully the cost-benefit analysis is justifiable.

References

1. Bellazzi, R., Ironi, L., Guglielmann, R., Stefanelli, M.: Qualitative models and fuzzy systems: an integrated approach for learning from data. Artif. Intell. Med. **14**, 5–28 (1998)
2. Christos, G., O'Flynn, B., Achille, K.: Automated zone-specific irrigation with wireless sensor/actuator network and adaptable decision support. Comput. Electron. Agric. **105**, 20–33 (2014)
3. Daniel Fisher, K., Hirut, K.: A low-cost microcontroller-based system to monitor crop temperature and water status. Comput. Electron. Agric. **74**, 168–173 (2010)
4. Digi International Inc.: XBee/XBee-Pro ZB OEM RF Modules. http://www.digi.com (2008)

5. Do Guen, Y., Ho Min, L., Ali, S., Joong Hoon, K.: Optimal pipe size design for looped irrigation water supply system using harmony search. Saemangeum Project Area. Sci. World J. Article ID 651763, 1–10 (2015)
6. Evans, R.G., Iversen, W.M., Kim, Y.: Integrated decision support: sensors networks and adaptive control for wireless site-specific sprinkler irrigation. Appl. Eng. Agric. **28**(3), 377–387 (2012)
7. Farid, T., Al-Hitmi, M., Benhmed, K., Tabish, R.: A fuzzy logic based irrigation system enhanced with wireless data logging applied to the state of Qatar. Comput. Electron. Agric. **98**, 233–241 (2013)
8. Genghuang, Y., Yuliang, L., Li, Z., Shigang, C., Qingguo, M., Hongda, C.: Automatic irrigation system based on wireless network. In: IEEE International Conference on Control and Automation Xiamen, vol. 1, pp. 2120–2125 (2010)
9. Guang-cheng, S., Ming-hui, W., Na, L., Min, Y., Prem, K., Dong-Li, S.: Growth and comprehensive quality index of tomato under rain shelters in response to different irrigation and drainage treatments. Sci. World J. Article ID 457937, 1–12 (2014)
10. Javadi Kia, P., Tabatabaee Far, A., Omid, M., Alimardani, R., Naderloo, L.: Intelligent control based fuzzy logic for automation of greenhouse irrigation system and evaluation in relation to conventional systems. World Appl. Sci. J. **6**(1), 16–23 (2009)
11. Joaquin, G., Villa-Medina, J.F., Nieto-Garibay, A., Porta-Gandara, M.A.: Automated irrigation system using a wireless sensor network and GPRS module. IEEE Trans. Instrum. Meas. **63**(1), 166–176 (2014)
12. Kati, M., Bruce, S., Jonathan, C., Frederick, D.: Plant response to evapotranspiration and soil water sensor irrigation scheduling methods for papaya production in south Florida. Agric. Water Manag. **97**(10), 1452–1460 (2010)
13. Mamdani, E., Assilian, S.: An experiment in linguistic synthesis with a fuzzy logic controller. Int. J. Man Mach. Stud. **7**, 1–13 (1975)
14. Rafael, M.C., Michael, D.: Automatic irrigation based on soil moisture for vegetable crops. Department of Agricultural and Biological Engineering, UF/IFAS, AE354 (2014)
15. Robert, W.C., Michael, J.D., Alan, B., Mark, H.: Wireless sensor network with irrigation valve control. Comput. Electron. Agric. **96**, 13–22 (2013)
16. Sabrine, K., El Houssaini, D., Jmal, M.W., Viehweger, C., Abid, M., Kanoun, O.: Precision irrigation based on wireless sensor network. IET Sci. Meas. Technol. **8**(3), 98–106 (2014)
17. Shaughnessy, S.A., Evett, S.R.: Canopy temperature based system effectively schedules and controls center pivot irrigation of cotton. Agric. Water Manag. **97**(9), 1310–1316 (2010)
18. Xiaohong, P., Guodong, L.: Intelligent water-saving irrigation system based on fuzzy control and wireless sensor network. In: Fourth International Conference on Digital Home, vol. 2, pp. 252–256 (2012)
19. Xin, D., Vuran, M.C., Irmak, S.: Autonomous precision agriculture through integration of wireless underground sensor networks with center pivot irrigation systems. Ad Hoc Netw. **11**(7), 1975–1987 (2013)
20. Yang, W., Boyd, S.: Fast model on control using online optimization. IEEE Trans. Control Syst. Technol. **18**(2), 267–278 (2010)

Artificial Neural Network Based Real-Time Urban Road Traffic State Estimation Framework

Ayalew Belay Habtie, Ajith Abraham and Dida Midekso

Abstract With the rapid increase of urban development and the surge in vehicle ownership, urban road transport problems like traffic accident and congestion caused huge waste of time, property damage and environmental pollution in recent years. To address these problems, use of information communication technology-based transport systems that can support maximum utilization of the existing road transport infrastructure has been proposed by different researchers. Road monitoring systems are one of these solutions which support road users to make informed decisions. However, the current road traffic monitoring systems use road side infrastructures for road traffic data collection and these technologies lack accurate and up-to-date traffic data covering the whole road network. By comparison, cellular networks are already widely deployed and can provide large road network coverage. Besides, 3G and 4G cellular networks provide mobile phone positioning facility with better performance accuracy and this opportunity can help to obtain accurate traffic flow information in cost effective manner on the entire road networks. The purpose of this chapter is to present our approach for real-time road traffic state estimation framework using the existing cellular network for road traffic data source and a neural network state estimation model. To evaluate the performance of the Artificial Neural Network model (ANN) both simulation and real world data is applied. The estimation accuracy using MAE and estimation availability indicated that reliable link speed estimation can be generated using this model and the estimated data can help to indicate real-time urban road traffic condition.

Keywords Cellular network · Positioning technology · Artificial Neural Network · Framework · State estimation

---

A.B. Habtie (✉) · D. Midekso
Department of Computer Science, Addis Ababa University, Addis Ababa, Ethiopia
e-mail: ayalew.belay@aau.edu.et

D. Midekso
e-mail: dida.midekso@aau.edu.et

A. Abraham
Machine Intelligence Research Labs (MIR Labs), Washington, DC 98071, USA
e-mail: ajith.abraham@ieee.org

© Springer International Publishing AG 2017
A. Abraham et al. (eds.), *Computational Intelligence in Wireless Sensor Networks*,
Studies in Computational Intelligence SCI 676, DOI 10.1007/978-3-319-47715-2_4

1 Introduction

Road traffic flow information is essential for the development of efficient traffic control and management systems. Real-time road traffic flow information is utilized for various purposes such as dynamic route guidance, incident detection, vehicle emission monitoring and is also useful for successful deployment of Intelligent Transport System (ITS) applications-Advanced Traffic Management Systems (ATMS) and Advanced Traveller Information Systems (ATIS). Particularly, ATMS and ATIS need accurate real-time road traffic flow information and short term traffic state estimation of the future for smooth traffic flow [1] and to deliver integrated road traffic flow information to road users.

Road traffic state estimation comprises different activities, road traffic data collection is the primary step. Traffic flow information can be obtained using different road traffic surveillance technologies. Fixed sensor technologies like inductive loop detectors, Video Image Processing (VIP), etc. are the current state-of-the-practice road traffic data collection tools to gather information about traffic flow in most part of the world [2]. But, these technologies don't cover wide scale of urban road network due to high deployment and maintenance cost. Moreover, with fixed sensors, it is possible to gather spot road traffic data which can't reflect road traffic flow of the entire road network.

Mobile probe-based road traffic surveillance technologies are alternatives to this luxury, fixed road side infrastructures. Mobile probe-based road traffic surveillance systems collect and process road traffic data locating the vehicle via GPS or mobile phones over the entire road network [3]. Floating vehicle technologies or dedicated vehicle probes are one category in mobile probe-based traffic monitoring systems which totally depend on GPS to localize the vehicle and its trajectory.

To have sufficient sensing capability utilizing floating vehicle technologies, significant percentage of vehicles should be equipped with GPS device and this mass deployment of GPS is expensive. Besides, automobile owners should also agree to share their location information to the traffic system. Furthermore, due to signal multipath and urban canyon obstruction, GPS doesn't work well in urban areas [4]. Hence, these traffic data collection technologies provide limited sample size [5] and the data collected can't be a representative for all vehicles on the entire road network [6].

Contrary to fixed sensor technologies and floating vehicle technologies, using the existing cellular network infrastructure to gather road traffic data offers large coverage capability as traffic data can be obtained continuously and also it is faster to set up, easier to install and needs less maintenance [7]. Here, vehicles on the road are assumed to have one or more turned on mobile phones (could be GPS-enabled or not) and information about the vehicle (location, speed, time-stamp) can be gathered using the positioning technology supported by the specific cellular network generation.

Mobile positioning technologies which aim to collect road traffic data in cellular networks can be either Handset-based or Network-based. In handset-based positioning methods, the mobile itself computes the position of the subscriber mobile phone and it is highly secured, can support gathering more measurements but demands high power consumption and also may need to incorporate special software and hardware in the handset as well as on the network [8]. Network-based positioning methods determine the location of mobile phones using the mobile network and services utilizing these positioning technology can support legacy mobile phones without upgrading, reduces power consumption, can initiate mobile phone positioning without any intervention but require software and hardware change on the infrastructure [8]. Individually, each of these positioning technologies can't provide wider area coverage and improved positioning accuracy simultaneously [9].

To improve positioning accuracy, coverage and communication latency of positioning technologies, combination of Handset-based and Network-based positioning techniques has been proposed [10, 11]. Combination of the measurements of these technologies is generally performed by either measure fusion algorithm or state vector algorithm. It was shown that measurement fusion algorithm outperforms state vector fusion algorithm particularly for tracking and localizing a moving vehicle [12].

In line with this, different researchers conducted field experiment to show the feasibility of using mobile phones as road traffic probes [13–15]. However, most of the experiments were done in estimating road traffic states on freeways and only few trials attempted monitoring traffic flow on arterial roads. According to Bacchus et al. [13] recommendation, future research effort should be done on gathering traffic data for arterials where no data is available. But road traffic state estimation on arterials is more difficult than estimation on freeways as arterials have low traffic volume, variable vehicle speed on different road networks and at road intersections it is common to get traffic lights for controlling purpose [14, 16]

Among previous deployments, most employed network-based mobile positioning methods which use network signal information like handoff, Angle of Arrival and Time of Arrival. Few of these deployments applied handset-based mobile positioning method to estimate road traffic flow. For example, Mobile Millennium [17] and Tao et al. [18] employed the handset-based (GPS-enabled mobile phone) method in experimenting the use of cellular network for road traffic flow estimation. From field evaluation, due to additional complexities of arterial vehicle roads, network-based methods can't provide sufficient amount of traffic data for traffic state estimation. Though GPS enabled mobile phones provided successful road traffic state estimation, additional communication cost and slow uptake of smart phones are mentioned as success obstacles [19]. But these can't be problems any more as UMTS networks have wide bandwidth for communication and according to Chi and Xavier [20], GPS-enabled Smartphone is expected to grow to 65.1 %, which is 2.6 billion units, in 2016 worldwide. Particularly in Africa and Middles East, GPS-enabled Smartphone sales volume rate rises to 56 % in 2013 where in Africa it is almost two times higher than the global average growth rate [21].

Once traffic data about vehicles on arterial roads is gathered and processed, the next step is the estimation of road traffic flow status for every road link on the entire road network. Over the years, variety of traffic state like travel time and traffic flow estimation models have been developed. But considering offline online applicability, existing road traffic state estimation techniques can be either model based or data driven based approach [22]. Although current practices on urban road traffic state estimation applied data driven approaches, no single predictor had yet been developed that presented itself to be universally accepted as the best, and proved to be a consistent and effective traffic state forecasting model for real-time traffic operation. But from all data driven traffic state estimation models, Artificial Neural Networks are very flexible in producing accurate multiple step-ahead estimation with less effort and are chosen to be best traffic modeling tool [23].

Recently, to overcome limitation of a single traffic state estimation model by advantage of another, hybrid models (combining parametric and non-parametric or model based and data driven method) drew much attention. For example, Yin et al. [24] developed fuzzy-neural model (FNM) to predict the traffic flows in an urban street network. Alescsandru and Ishak [25] also presented a model-based and memory-based hybrid system to improve performance of freeway speed forecasting systems. Zheng et al. [26] developed a Combined Neural Network Model (CNNM) for short-term freeway traffic flow prediction and Van Lint et al. [27] applied a Kalman filter neural network to forecast short-term travel time on freeways. Anderson and Bell [28] used the model based microscopic traffic flow VISSIM with Neural Network for traffic state estimation techniques and a queuing model for travel-time prediction in urban road networks. The work of Tao et al. [18] also presented the use of Kalman filtering integrated with a microscopic simulation model SUMO in urban traffic state estimation using A-GPS based vehicle location data.

In this chapter a real-time road traffic state estimation framework that utilizes the existing cellular network infrastructure for road traffic data collection with a neural network model estimator is discussed. The framework integrates different modules at which different models are proposed to be used in the process of traffic state estimation. To evaluate the framework, a hybrid method of combining a three-layer Artificial Neural Network model and the microscopic simulation model SUMO is used. The chapter describes the proposed framework and its applications in detail based on our work [29–32].

The rest of this chapter is organized as follows: Sect. 2 describes the proposed road traffic state estimation system. Section 3 discusses the framework application, Sect. 4 presents the road link estimates and the summary is depicted in Sect. 5.

2 Road Traffic State Estimation System

The core milestones of real-time Road Traffic State Estimation based on cellular network signaling include location data collection, Mobile phone mobility classification, map matching and route determination, road traffic condition estimation and dissemination of traffic information to road users [32]. These activities are comprised in a three layer architecture: data collection layer, traffic state estimation layer and database server layer [2, 33, 34]. The data collection layer is responsible to collect real-time location data of vehicles on the road and send the data to the next layer-

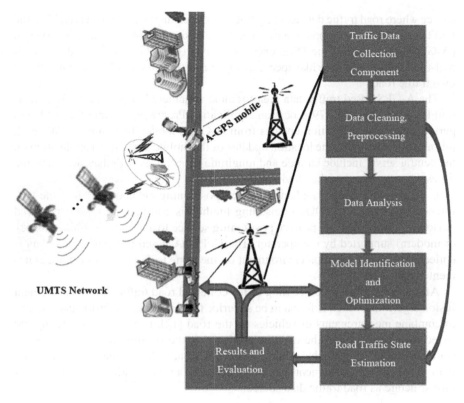

Fig. 1 General framework for short term arterial traffic state estimation

estimation server. The road traffic state estimation layer does the data preprocessing, updates data to the database server, estimates real-time traffic flow and provide the information to road users on mobiles or on to the Internet. The third layer, database server, stores digital map which is used for road traffic state estimation and information dissemination processes.

Considering the detail activities we have done in our work [29] while using the cellular network for real-time road traffic data collection, the proposed real-time road traffic state estimation framework is depicted in Fig. 1. The proposed framework, depicted in Fig. 1, has five basic components.

2.1 Traffic Data Collection Component

This component represents the data sources which provide real-time road traffic measurements on the basis of which real-time road traffic flow condition is estimated. As it is indicated in the framework, the cellular network is used as road traffic data source where road traffic data can be gathered. For example, in our research [29], the UMTS cellular network positioning standards, Assisted Global Positioning System (A-GPS) and Uplink Time Difference of Arrival (U-TDOA) were used to gather real-time road traffic data like speed, location and time-stamp of a moving vehicle on a traffic road.

The A-GPS based traffic data measurement is gathered utilizing Java Specification Request 179 (JSR 179) Location Application Programming Interface (API) to periodically request location updates from A-GPS mobile phone moving along all journey of the vehicle. The location updates of the mobile sent to java application on the central server include latitude and longitude coordinates with their accuracy, the timestamp and speed of the moving vehicle.

Unlike A-GPS which is a handset based positioning method, Uplink Time Difference Of Arrival (U-TDOA) poisoning method is considered as network based poisoning system that can provide positioning service to all kind of mobiles (legacy or modern) supported by the operator. The U-TDOA based moving vehicle data is collected using a simulation environment on the same traffic road A-GPS measurement is gathered.

After cleaning and preprocessing of the collected road traffic data, measurement fusion algorithm, which is found to be superior from state vector fusion particularly to combine measurements of vehicles on the road [12], is employed to get hybrid based measurement and the experiment done on the positioning accuracy of the cellular network revealed that reliable road traffic data with acceptable accuracy (i.e., with in the E-911 requirement) can be gathered using the existing cellular network infrastructure as road traffic data source.

2.2 Data Cleaning and Preprocessing

Usually, the raw traffic sensor data is not good enough to use it directly for road traffic management activity. Primarily data cleaning activities enable to remove abnormal data including abnormal velocity, wrong direction and time duplication. Then other preprocessing activities like data checking, data completion and data correction will be performed [34]. Besides, when In-vehicle mobile phones are used as traffic probes, activities like mobility classifications and map matching are done during preprocessing. For example, in our work [29] where traffic data is collected based on the UMTS cellular network positioning method of A-GPS and U-TDOA, preprocessing activities like coordinate transformation of A-GPS measurements, linearization of U-DOA based traffic data, classification, map matching and fusion of the measurements were done as it is depicted in Fig. 2.

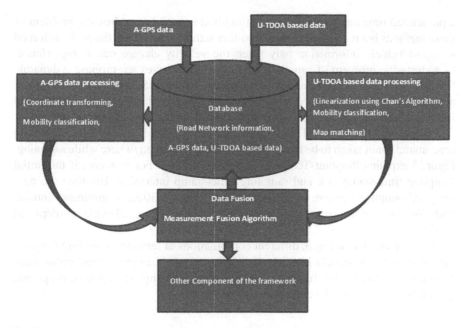

Fig. 2 Preprocessing activities on fusion based A-GPS, U-TDOA traffic data

2.3 Data Analysis

This component of the framework represents the statistical activities performed for data sampling from raw traffic sensor data which was cleaned and preprocessed using appropriate techniques/procedures. For acceptable quality of urban road traffic state estimation with probe based monitoring, the percentage of vehicles/mobile serving as traffic probe, i.e. penetration rate is with high concern [35, 36]. Accordingly, mobile probe based urban freeway traffic monitoring can provide sufficient data when the penetration rate is from 3 to 5 %. Though the minimum sample size of probe vehicles is dependent to factors like road type, link length and sample frequency [35], for reliable speed estimation of arterial roads, a minimum penetration rate of 7 %, i.e. at least 10 probe vehicles traversing a road section (every road link) successfully is required [35–37].

The other crucial issue in road traffic state estimation using mobile phones as traffic probes is the sampling frequency. For example, A-GPS mobile phones receive location updates at every 1–3 s and collection of data from large number of mobile probes may cause network carrier congestion. To solve this problem three approaches are proposed: temporal (report vehicle information at prescribed time), spatial (probes send information when passing predefined location) and pinpoint method (report based on vehicle velocity change rate) [33].

The above sampling frequency approaches, temporal and spatial can't necessarily help to collect useful data for traffic state estimation as the collection is dependent on a predefined time and location Pinpoint method is useful in addressing problem of these methods but it may not enable to collect sample size data as the probe is forced to report vehicle information only when the velocity change rate is big. Hence, to address the problem of sampling frequency approaches we propose a dynamic "Pinpoint-Temporal" sampling frequency method which is a combination of Pinpoint and Temporal. In this method the required sample size data incorporating vehicle information at change of state points with variable initial time-stamp is collected.

During sampling, to avoid road traffic state estimation biases, the initial sampling time-stamp is not taken to be fixed as the vehicle could be anywhere while sampling. Figure 3 explains Pinpoint-Temporal sampling strategy. For instance, if the initial sampling time-stamp is i, and sampling time-stamp interval is 10 s then the next temporal sampling frequencies will be $i + 10, i + 20, i + 30$ etc. Combining Pinpoint with this temporal strategy, the points where the data recording takes place is depicted in Fig. 3.

As it is indicated in Fig. 3, different combinations of temporal sampling frequencies with different initials like $i, i + 1, i + 2$, etc. could improve road traffic state estimation accuracy [38]. In the process, the vehicle change state points, p_1, p_2 and p_3 unless repeated, will be considered in the sample.

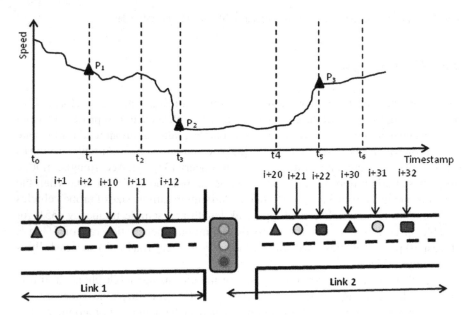

Fig. 3 Pinpoint-temporal sampling based vehicle information recording at different links

2.4　Model Identification and Optimization

This component of the proposed real-time road traffic state estimation framework represents appropriate approaches used to estimate road traffic sate with acceptable accuracy. Current practices of road traffic state estimation techniques applied either model based or data driven approaches [22]. Model based traffic state estimation approaches use the analytical traffic model Lighthill–Whitham–Richards (LWR) model or simulation based models which are more suitable to model complex road traffic flows [40]. The data driven approaches are data intensive and commonly applied for real-time traffic state estimations. These models include statistical parametric and non-parametric techniques [23]. Different studies demonstrate non-parametric real-time traffic state estimation techniques performing well due to the ability to capture non deterministic and complex non linearity of traffic flow series [41]. Particularly, Artificial Neural Networks have been applied extensively in short term traffic forecasting field and acknowledged to be a promising approach because of its superiority in modeling complex nonlinear relationships [42, 43]. However, no single predictor had yet been developed that presented itself to be universally accepted as the best and effective traffic state forecasting model for real-time traffic operation [44]. Hence, selection of appropriate methodological approach is a major issue in traffic state estimation. In our work [45], hybrid method of combining Artificial Neural Network (data driven approach) and the microscopic simulation model (model based approach) is found to be better when mobile phones are used as traffic probes.

Artificial Neural Network Model for Urban Arterial Road Traffic State Estimation Different types of artificial neural networks (ANN) have been proposed in the past few years for estimation purpose. The most popular connected multilayer perceptron (MLP) neural network architecture is chosen in this study as it is extensively applied in transportation applications, has very good capability, and easy to implement so long as there are enough neurons in the hidden layer [46]. The architecture of ANN is composed of set of nodes and connections organized in layers. In this work a three layer ANN is used: input layer, hidden layer and output layer. The input layer enables to receive external information and the output layer is the layer where problem solution is acquired. Usually one or two hidden layers are used between the aforementioned layers but single hidden layer is proved to be enough for ANNs to estimate nonlinear functions [46]. It is the hidden nodes in the hidden layer that allow the neural network to detect the feature to capture the pattern in the data, to perform nonlinear mapping between input and output variables.

As it is discussed in Sect. 2.1, the cellular network is used as road traffic data source and the data collected based on mobile phones as probes contain vehicle position, time stamp and vehicle speed on the road link. Hence position, time stamp and speeds are used as input data in the ANN model. The structure of the ANN model, which is experimentally proved with different traffic demand including 20 %

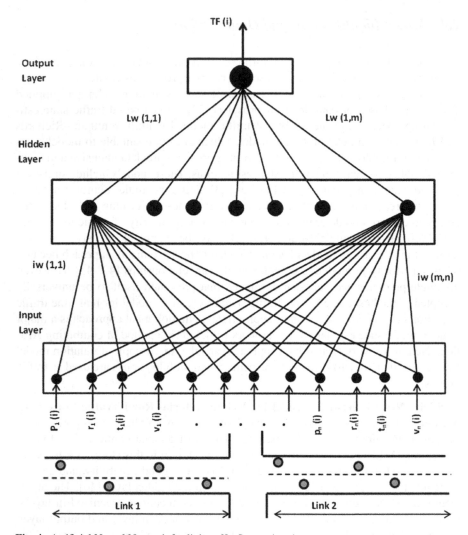

Fig. 4 Artificial Neural Network for link traffic flow estimation

demand increase, 50 % demand increase and 100 % demand increase, is adapted from Zheng and Zuylen [39] as shown in Fig. 4. According to this author, the accuracy of the ANN model is sensitive to position input whereas the speed input is not crucial factor influencing the ANN performance.

The mathematical model for the input layer, hidden layer and output layer is given as follows.

Input layer

$$
x(i) = \begin{bmatrix} x_1(i) \\ \vdots \\ x_n(i) \end{bmatrix} = \begin{bmatrix} p(i) \\ r(i) \\ t(i) \\ v(i) \end{bmatrix}, p(i) = \begin{bmatrix} p_1(i) \\ \vdots \\ p_n(i) \end{bmatrix}, r(i) = \begin{bmatrix} r_1(i) \\ \vdots \\ r_n(i) \end{bmatrix},
$$

$$
t(i) = \begin{bmatrix} t_1(i) \\ \vdots \\ t_n(i) \end{bmatrix}, v(i) = \begin{bmatrix} v_1(i) \\ \vdots \\ v_n(i) \end{bmatrix}, \tag{1}
$$

where p(i) is position vector, r(i) is link id vector, t(i) is time stamp vector and v(i) is speed vector

Hidden layer

$$
H(i) = \begin{bmatrix} h_1(i) \\ \vdots \\ h_m(i) \end{bmatrix} = \begin{bmatrix} \varphi(\sum_{j=1}^{N} w_{j,1} x_j(i) + b_1) \\ \vdots \\ \varphi(\sum_{j=1}^{N} w_{j,m} x_j(i) + b_m) \end{bmatrix}, \tag{2}
$$

where $h_m(i)$ denotes the value of the mth hidden neuron, $w_{j,m}$ represent the weight connecting the jth input neuron and the mth hidden neuron, b_m is bias with fixed value for the mth hidden neuron and φ is the transfer function.

Output layer

$$
y(i) = TF(i) = \varphi(\sum_{k=1}^{m} w_k h_k(i) + b), \tag{3}
$$

where $y(i)$ and $TF(i)$ denote estimated traffic speed of probe vehicle i on the link under consideration, w_k represent the weight connecting the kth hidden neuron and the output neuron, b is bias for the output and φ is the transfer function.

2.5 Road Traffic State Estimation

This component of short term arterial traffic state estimation framework represents techniques/procedures employed to estimate traffic flow at every link of vehicles route. Theoretically, road traffic state can be calculated without using any estimation models when all the vehicles travelling on the road are used to collect traffic data [47]. However, it is difficult due to some factors like public's awareness of privacy protection [48] and some information sharing problems [49].

Road traffic state estimation is performed based on transforming criteria (methods) on the calculated average speed of vehicles on the road links [47]. Basically there are three ways of calculating road traffic speed of vehicles: Average Method, Integral Method and Fitting Method.

Average Method (AM) This method is the simplest one where road traffic speed of vehicles is calculated using arithmetic mean of vehicle location points [49]. For a set of vehicle localization points with speed($V_t^r = \{v_{i,t}^r\}$, $(i = 1, 2 \ldots .n)$), on segment r at an interval time (t), the average road traffic speed is given by:

$$V_t^r = avg_{i=1}^n(v_{i,t}^r) \tag{4}$$

To avoid biases, the data need to satisfy requirements of statistics. The average method has been applied in different studies for road traffic state estimation activity, Tao et al. [18], Chen et al. [50] are some of them.

Integral Method (IM) This method determines traffic speed as traveled distance (D) divided by the time (T) spent on the distance. If a vehicle (m) is traveling on a road segment (r) at time interval (t), the vehicle mean speed using integral is given as:

$$V_t^{r,m} = D/T = \int_t^{t+1} vdt/T \tag{5}$$

According to Quiroga [51], it may be necessary to approximate equation (5) as speed and time of a vehicle traveling on a road will not have relationship and the approximation is given as:

$$v_t^{r,m} \approx \frac{1}{T}\{v_{0,t}^{r,m}(\frac{t_1 - t_0}{2}) + \sum_{n-1}^{N-1} v_{n,t}^{r,m}(\frac{t_{n+1} - t_{n-1}}{2}) + v_{N-1,t}^{r,m}(\frac{t_N - t_{N-1}}{2})\} \tag{6}$$

where n is the point index of one vehicle, if multiple vehicles (m > 1) are passing on the segment at time t the calculation is done by averaging $v_t^{r,m}$.

Fitting Method (FM) This method uses the position information of the vehicle. According to Kong et al. [52], if L is the total length of road r, the speed of a vehicle during an interval time (T) on the whole segment is given by:

$$v_t^r = \left[\int_{(k-1)T}^{KT} v_t^r dt\right]/T \tag{7}$$

Most studies, for example Qiankun et al. [53], Kong et al. [54], Tao et al. [18], which were conducted on road traffic state estimation employed the Average Method (AM) to estimate traffic flow on road links. However, an experimental evaluation on the performance of these methods revealed that as the number of vehicles on arterial road becomes seven or more, the performance of Integral Method (IM) is better than the other two [47]. Thus, we propose to use IM method in the road traffic state estimation process.

2.6 Results and Evaluation

This component of the framework evaluates and disseminates road traffic flow information to road users on their mobile phones or on the Internet for universal access. During evaluation, comparison of road traffic state estimation model result and ground truth based traffic state data is done and if accuracy and coverage of estimation result is not acceptable, optimization of the identified road traffic state estimation framework is done with different techniques like improved probe penetration rate and sampling frequency [18], special point elimination method, dynamic boundary method and IM–FM fusion method [47]. However, evaluation of the results from field tests is not suitable for statistical analysis particularly for arterial roads [18]. Hence simulation based individual vehicle tracks and aggregate traffic states can be used as ground truth during evaluation [15].

3 Framework Application

In Sect. 2.1, it was discussed that the UMTS cellular network is used as road traffic data source. For the application of the framework, real-world data was gathered based on a java application utilizing JSR 179 API on A-GPS mobile phone. But, due to security reason, it was impossible to get U-TDOA based traffic measurements on a traffic road. Hence, to model the arterial road traffic, a microscopic simulation package "Simulation of Urban Mobility" (SUMO) [55, 56] is employed. The SUMO packages NETCONVERT and DFROUTER are used to generate road network and vehicle route. From simulation, aggregated speed information for each road link named "aggregate edge states" and floating car data (FCD) export file are generated for further analysis. The information on aggregate edge states include road edge IDs, time intervals, mean speeds, etc. The aggregate speeds are used to determine ground truth of traffic flow. Whereas the FCD output file records the location coordinates of every moving vehicle on the sample road at every time-stamp, vehicle speed, edge IDs and this data file is used as simulation for in-vehicle mobile based traffic information system.

The generated vehicle location data is firstly preprocessed to reflect "realistic" location sample that A-GPS mobiles may provide in real world situation. Next, Data analysis activities like penetration rate, sampling frequency algorithms are employed on the preprocessed location data. Using appropriate proportion of the data, training of Artificial Neural Network (ANN) model of the selected road network is done. Then speed estimates of the trained ANN are allocated to each of the sample road links and aggregated on predefined time interval. For performance assessment of the identified state estimation model, accuracy of speed estimation is evaluated using Mean Absolute Error (MAE) as well as comparison with the "ground truth" speed value of SUMO aggregate state. Finally, estimated speed values are classified in to different traffic flow conditions using threshold techniques and presented as colored road segments on road users' mobile phones or on the Internet.

3.1 Test Urban Road Network

As it is shown in Fig. 5, part of Addis Ababa city road network is chosen as simulation case study. The OpenStreetMap (OSM) xml file of the selected road network is edited using Java OpenStretMap (JOSM) [57] to remove road edges that can't be used by vehicles like road ways to pedestrian etc., and also for simplicity all road edges are set to one-way.

The simplified form of the road network consists of 13 nodes, 12 links with length ranging from 169 to 593 m and 4 traffic lights as shown in Fig. 6. On SUMO command window, network file encoding traffic light logic, speed limit of road links and road link priorities are generated from the OSM file using NETCONVERT.

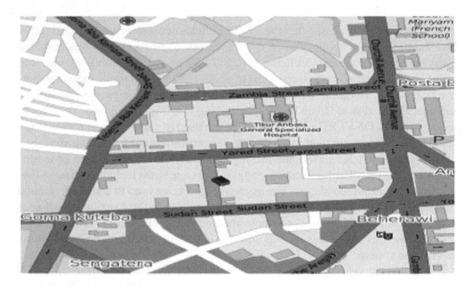

Fig. 5 Sample road network from OpenStreetMap

Fig. 6 Simplified diagram from SUMO

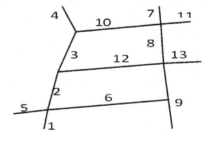

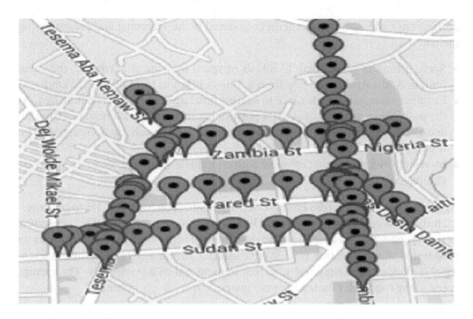

Fig. 7 Real A-GPS based moving vehicle location data

Then using DUAROUTER random routes of vehicles traveling on the road network, number of vehicles to be emitted to the road, start and end of traffic flow are determined. These randomly generated vehicles and their routes are run on SUMO and as a result of 3600-s interval, 718 probe vehicles are generated with random trips. In order to mimic the real traffic situation on this road, free flow traffic demand is used with maximum speed 30 km/h, which is the speed limit in the real situation. Besides, the data sets from the simulated network, real data set was collected by a car with A-GPS device driving on the sample road networks and location updates in terms of longitude and latitude, time stamp and speed were recorded for every 3 s. The real A-GPS location data collected within 45 min is depicted in Fig. 7.

3.2 Data Preparation

The SUMO simulation is conducted for 1-h without incident. Data from the first 15 s of simulation were considered to be warm-up period and were not used in the analysis. Every second, position, time stamp and speeds of vehicles were recorded. The output files "aggregate edge states" and FCD output, before used for further analysis, data screening and preprocessing is performed.

Data Screening and Preprocessing: The first challenge in using the cellular network based mobile probe system is distinguishing non-valid probes (pedestrians etc.). But probe data in this system come from our service subscribers and validity of traffic probes is not an issue. Other criteria considered in the screening and preprocessing include:

- Speed estimates greater than 13.89 m/s of speed limit are eliminated as the speed limit in Addis Ababa city vary from 8.3 m/s (30 km/h) to 13.89 m/s (53 km/h).
- Location estimates with distance to nearest link larger than 20 m is eliminated. As the positioning accuracy of A-GPS positioning method is maximum of 20 m [58] and also to differentiate closely spaced parallel urban roads, an accuracy of 20 m is expected [59].
- In-vehicle mobile location estimate with coordinate (0, 0) and with zero speed are eliminated.

Aggregation of link speeds: From SUMO simulation conducted for 1 h, to character-ize traffic flow condition on every road link of the sample road networks, aggregation of link speed estimates over a specific time interval is performed. Previous works on road traffic state estimation [13, 35] discussed a 10 min aggregation time is a rea-sonable choice considering real-time requirement and data availability. The average speed along road link r during the time interval $[t_k, t_{k+\Delta T}]$ is given as:

$$V_{av}^r\left(t_k, t_{k+\Delta T}\right) = \frac{1}{n_{t_k, t_{k+\Delta T}}^r} \sum_{k=t_k}^{t_k+\Delta T} v_r(k) \qquad (8)$$

where $v_r(k)$ is the available speed estimate on link r and $\frac{1}{n_{t_k, t_{k+\Delta T}}^r}$ is total number of available speed estimates. The mean traveling speed of the road links on a 10-min time interval is recorded in the "aggregated edge states "and it is depicted in Fig. 8. From the figure, all road links except link 1 and link 5 are expected to have smooth traffic flow as their speed is above 7 m/s (link speed limit is 30 km/h). However, link 5 has medium traffic flow and link 1 is with medium traffic flow from 10 to 20 and 50 to 60 min. From 20 to 50 min the traffic flow on this road link is improved.

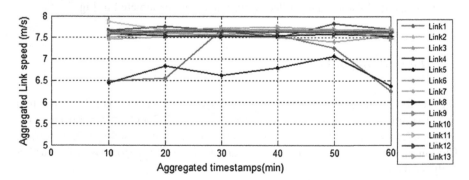

Fig. 8 Aggregated road link speeds—"ground truth speed"

Data Sampling Process From the 1 h SUMO simulation, the simulation output file, FCD output, generates large amount of simulated mobile probes at real time. At every second, location updates (in terms of x, y or longitude, latitude) for the mobile probes are recorded. There are in total 437 probes and the time they spent range from 10 to 202 s. Figure 9 shows location data collected from these probes, which are aggregated in 10 min. The FCD output file contains detail information of each vehicle/mobile and grows extremely large. Hence, converting this location data in to more compressed one is necessary [18]. Accordingly, to degrade location data, one can set up a specified percentage of simulated vehicles/mobiles to be traffic probe. As it is discussed in Sect. 2.3, previous studies suggest that for arterial roads reliable speed estimation, a minimum penetration rate of 7 %, i.e. at least 10 probe vehicles traversing a road section (every road link) successfully is required [35–37] although factors like road type, link length and sample frequency affect the minimum sample size.

In this research work, the sample is taken considering the road link length as it is indicated in Table 1 and sampling frequency is based on "Pinpoint-Temporal" which is elaborated in Sect. 2.3.

As it is presented in Table 1, for road link length 300 m and below 10 probes are taken (which is the minimum recommended in the literature), 300 to 600 m 15 probes and more than 600 m link length 20 probes and totally 165 probes (37.6 %) are considered in the sample. The sampling frequency employed is "Pinpoint-Temporal" with 10 s time interval and 15 location updates of each probe is collected including

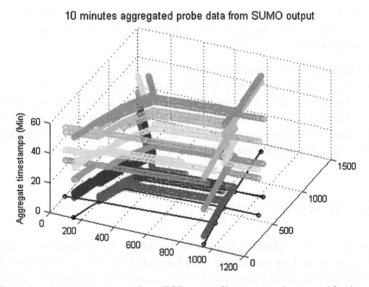

Fig. 9 Probe location data generated from FCD output file aggregated at every 10-min

Table 1 Number of probes taken for the sample from the FCD output based on road link length

Link # (street name)	Link length (m)	Number of sample probes	Number of location updates per probe
1 (Tesema Aba Wekaw Street)	274.6	10	15
2 (Tesema Aba Wekaw Street)	233.4	10	15
3 (Tesema Aba Wekaw Street)	258.8	10	15
4 (Tesema Aba Wekaw Street)	194.2	10	15
5 (Sudan Street)	194.2	10	15
6 (Sudan Street)	695.8	20	15
7 (Churchill Avenue)	593.9	15	15
8 (Churchill Avenue)	233.3	10	15
9 (Churchill Avenue)	531.4	15	15
10 (Zambia Street)	484.3	15	15
11(Nigeria Street)	69.5	10	15
12 (Yared Street)	601.7	20	15
13 (Ras Damtew Street)	214.6	10	15

the first and last probe information as the proposed state estimation method, which is discussed in Sect. 2.5 is integral method (IM). Figure 10 plots location data of the sample probes aggregated in 10 min.

Data for Training and Evaluation After the sampling process, the extracted data were used for training the neural network and estimating the link travel speed. A total of 165 probe data were simulated and using random approach of dividing the available dataset for ANN development [60], 110 probe data (two-third of the data) were used for training process and the other 55 probe data (one-third of the data) were used for performance evaluation.

3.3   Neural Network Training

A training process is needed before the ANN model can be applied to estimate traffic state. In the process, three procedures including training, testing and validation were conducted. The total training data set (110 probe data) were divided in to three subsets [61] which are 88 probe data (80%) for training, 11 probe data (10%) for testing and 11 probe data (10%) for validation. During the training process, different hidden neurons like 10, 15, 20, 25 were chosen. During testing the performance in terms of

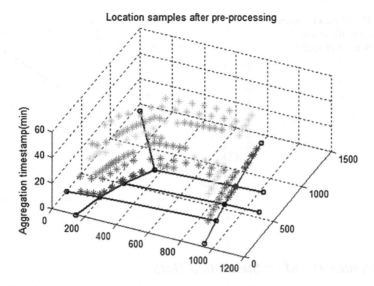

Fig. 10 Sample probe location data aggregated at every 10-min

Mean square error (MSE) for the case of 10, 15, 20 and 25 neurons is compared and 15 hidden neurons were used to build the network. Levenberg-Marquardt algorithm (Trainlm) [62] was chosen so that the over fitting phenomenon can be avoided. Moreover, the algorithm can provide fast convergence even for large networks with few hundred weights. The trained ANN model is applied to estimate link traffic state under free flow but proved even in over saturated condition [38].

3.4 Evaluation

To evaluate how the ANN model performs, the performance indicators Root Mean Square Error (RMSE) and Mean Absolute Percentage Error (MAPE) are used and defined as follows.

$$RMSE = \sqrt{\frac{1}{n} \sum_{k=1}^{n} (v_{pv,k} - v_{true,k})^2}, \tag{9}$$

$$MAPE = 100 * \frac{1}{n} \sum_{k=1}^{n} \left| \frac{v_{pv,k} - v_{true,k}}{v_{true,k}} \right|, \tag{10}$$

where $v_{pv,k}$ is the estimated travel speed of the kth probe vehicle and $v_{true,k}$ is the true link speed of the kth probe vehicle recorded by data collection points.

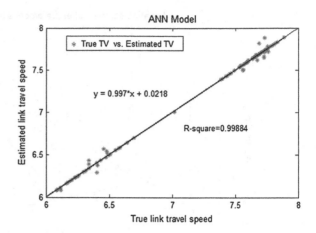

Fig. 11 Correlation between estimated link travel speed and true link travel speed

3.5   Results Based on Simulation Data

The trained ANN model is used to estimate link travel speed with a simulation data input. A correlation between the estimated link travel speed based on ANN model and the true link travel speed which is computed from the simulation data using the integration method (IM) of calculating vehicle speed is performed. As it is depicted in Fig. 11, the estimated link travel speed has very high correlation with the true link travel speed ($R^2 > 99\%$). The linear regression between the estimated and true (simulated) link speed that predicts the best performance among these values has an equation $y = 0.997*x + 0$ indicating the trained ANN model performs reasonably well, where x represents true speed and y estimated link travel speed.

The performance of the estimation method in terms of RMSE and MAPE is 0.029325 and 0.127 % respectively with average speed of 7.191 m/s.

3.6   Results Based on Real A-GPS Data

The trained ANN model was also applied to estimate travel speed based on real A-GPS data. A car with A-GPS based mobile phone traveled on the sample road network and location updates in terms of longitude, latitude, time-stamp, speed and accuracy is recorded at every 3 s for about 45 min (see Fig. 7). A sample using "Pinpoint-Temporal" method with 10 s time interval is taken and at every road link, 15 location points i.e. total of 195 A-GPS based vehicle location points are taken. The estimation result is shown in Fig. 12. Each point represents travel speed on each trip i.e. at every road link. From the regression formula in the figure, it can be seen that the trained ANN model performs very good. The RMSE and MAPE are 0.101034 and 1.1877 % respectively which show the possible application of the ANN model to real link travel speed measurements.

Fig. 12 Correlation between
estimated link travel speed
and true link travel speed
using real A-GPS data

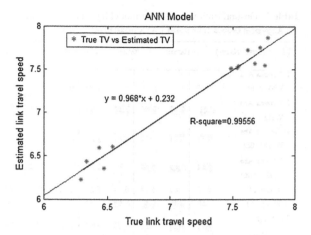

4   State Estimates of Road Links

As it is presented in Sects. 3.5 and 3.6, the trained ANN model provides a very good
link travel speed estimation using both simulated as well as real A-GPS data. In
this section, the trained ANN model is applied to estimate sample road link states
using link travel speeds and a comparison is made against the ground truth speed
estimates named as "aggregated edge states". For this analysis, a sample of 195 probe
vehicle points are taken from SUMO FCD output file based on "Pinpoint-Temporal"
sampling frequency method with a 10 min time interval. The result of the analysis is
shown in Table 2.

Two performance indicators Mean Absolute Error (MAE) and Estimation Availability are used to evaluate the ANN model estimation accuracy and system coverage
respectively. MAE is defined as mean of absolute difference between the ground truth
speed on a link and the estimated speed. The estimation speed availability is the fraction of links having speed estimation in the time interval considered.

Considering the state-of-the-art traffic speed classification in urban areas [50],
speed thresholds are employed to classify the estimated road traveling speed. Accordingly, three traffic condition levels: Green, Yellow and Red where green level (smooth
traffic) if link speed is above 7 m/s, yellow level (medium traffic condition) when
link speed is between 4 and 7 m/s and red level (congested traffic condition) when
link speed is below 4 m/s are used as it is shown in Table 2. From the table, it is
shown that medium traffic condition is detected on link 5 and partly on link 1. The
average estimation accuracy based on MAE is 0.28 m/s and all sample road networks
have link speed estimation using the trained ANN model. Moreover, these estimated
traffic conditions will be color-coded on the road network and presented on road
users' mobile display for real-time analysis.

Table 2 Ground truth versus ANN model based average link speeds

Link # (Street Name)	Aggregated Ground Truth Speed	ANN model based Speed Estimates with 10min Interval											MAE
	0-10min		10-20min		20-30min		30-40min		40-50min		50-60min		
1(Tesema Aba Wekaw St)	6.48	6.20	6.56	6.91	7.69	7.01	7.56	7.43	7.24	7.07	6.26	6.29	0.28
2(Tesema Aba Wekaw St.)	7.61	7.37	7.60	7.82	7.55	7.31	7.64	7.53	7.59	7.99	7.60	7.56	0.21
3(Tesema Aba Wekaw St.)	7.50	7.11	7.60	7.91	7.57	7.15	7.68	7.69	7.65	7.66	7.55	7.20	0.23
4(Tesema Aba Wekaw St.)	7.67	7.63	7.77	7.53	7.67	7.48	7.50	8.35	7.84	8.02	7.60	7.82	0.23
5 (Sudan St.)	6.45	6.62	6.84	6.69	6.63	6.02	6.80	6.61	6.98	6.67	6.38	6.89	0.32
6(Sudan St.)	7.59	7.28	7.03	7.91	7.07	7.98	7.63	7.27	7.08	7.48	7.04	8.09	0.33
7 (Churchill Avenue)	7.47	7.09	7.52	7.11	7.53	7.52	7.50	7.79	7.41	7.98	7.53	7.84	0.32
8 (Churchill Avenue)	7.59	7.49	7.55	7.07	7.54	7.08	7.55	7.33	7.59	7.56	7.54	7.84	0.26
9 (Churchill Avenue)	7.66	8.06	7.67	7.58	7.68	8.07	7.70	7.77	7.67	7.78	7.48	7.33	0.21
10 (Zambia St.)	7.57	7.07	7.64	7.35	7.62	7.36	7.62	7.97	7.62	7.55	7.62	7.86	0.29
11(Nigeria St.)	7.87	7.81	7.68	7.69	7.72	7.32	7.64	7.55	7.65	7.91	7.71	7.93	0.14
12 (Yared St.)	7.65	7.41	7.62	7.05	7.65	7.85	7.62	7.81	7.65	7.68	7.64	7.80	0.23
13(Ras Damtew St)	7.61	7.88	7.69	7.49	7.72	8.00	7.75	7.72	7.70	7.82	7.68	7.52	0.18

5 Summary

In this chapter, a method of real-time urban traffic state estimation is presented. A three-layer Artificial Neural Network model is proposed to estimate complete link traffic state. The inputs to the ANN model include probe vehicle's position, time stamps and speeds. Based on the microscopic traffic simulation SUMO, "aggregate edge state" which is the ground truth link travel speed and FCD output data is generated. From the FCD output, a sample based on "Pinpoint-Temporal" method is extracted and the dataset is divided to train as well as evaluate the ANN model. Besides, real A-GPS data gathered using A-GPS mobile phone on a moving vehicle on the sample roads is used to evaluate the ANN model. The performance of the ANN model is evaluated using the performance indicators RMSE and MPAE and on average, the MPAE is less than 1.2 %.

The trained ANN model is also used to estimate the sample road link speeds and compared with ground truth speed (aggregate edge states) on a 10-min interval for 1 h. The estimation accuracy using MAE and estimation availability indicated that reliable link speed estimation can be generated and used to indicate real-time urban road traffic condition.

References

1. Yang, Z., Bing, Q., Lin, C., Yang, N., Mei, D.: Research on short-term traffic flow prediction method based on similarity search of time series. Math. Probl. Eng. 1–9 (2014)
2. Rewadkar, D., Dixit, T.: Review of Different Methods Used for Large-Scale Urban Road Networks Traffic State Estimation. J. Emerg. Technol. Adv. Eng. 3(10), 369–373 (2013)
3. Leduc, G.: Road traffic data: Collection methods and applications. Work. Pap. Energy Transp. Clim. Change 1, 55 (2008)
4. Calabrese, F., Colonna, M., Lovisolo, P., Parata, D., Ratti, C.: Real-time urban monitoring using cell phones: A case study in Rome. IEEE Trans. Intell. Transp. Syst. 12, 141–151 (2011)
5. Pueboobpaphan, R. and Nakatsuji, T.: Real-time traffic state estimation on urban road network: the application of unscented Kalman filter. In Proceedings of the Ninth International Conference on Applications of Advanced Technology in Transportation, pp. 542–547 (2006)
6. Caceres, N., Wideberg, J., Benitez, F.: Review of traffic data estimations extracted from cellular networks. IET Intell. Transp. Syst. 2, 179–192 (2008)
7. Tao, S., Rodriguez, S., Rusu, A.: Vehicle location using wireless wide area network. In: 2010 Third Joint IFIP Wireless and Mobile Networking Conference (WMNC), pp. 1–6 (2010)
8. Lovisolo, P., Dario, P. and Carlo. R.: Real-Time Urban Monitoring Using Cellular Phones: a Case-Study in Rome. IEEE Trans. Intell. Transp. Syst. 21(1) (March 2011), 141–151 (2007)
9. Bensky, A.: Wireless Positioning Technologies and Applications. Artech House, Norwood (2007)
10. Tao, S., Manolopoulos, V., Rodriguez, S., Ismail, M. and Rusu, A.: Hybrid vehicle positioning and tracking using mobile phones. In: 2011 11th International Conference on ITS Telecommunications (ITST), pp. 315–320 (2011)
11. Abo-Zahhad, M., Ahmed, S.M., Mourad, M.: Hybrid Uplink-Time Difference of Arrival and Assisted-GPS Positioning Technique. Int. J. Commun. Netw. Syst. Sci. 5, 303–312 (2012)
12. Habtie, A.B., Ajith, A. and Dida. M.: Comparing Measurement and State Vector Data Fusion Algorithms for Mobile Phone Tracking Using A-GPS and U-TDOA Measurements. Hybrid Artifi. Intell. Syst. 9121, 592–604 (2015)
13. Bacchus, M.A., Hellinga, B. and Izadpanah, M.P.: An Opportunity Assessment of Wireless Monitoring of Network-Wide Road Traffic Conditions. Dept. Civil Eng., Univ. Waterloo, Waterloo, ON, Canada (2007)
14. Fontaine, M.D., Smith, B.L., Hendricks, A.R., Scherer, W.T.: Wireless location technology-based traffic monitoring: preliminary recommendations to transportation agencies based on synthesis of experience and simulation results. Trans. Res. Rec.: J. Transp. Res. Board 1993, 51–58 (2007)
15. Bar-Gera, H.: Evaluation of a cellular phone-based system for measurements of traffic speeds and travel times: A case study from Israel. Transp. Res. Part C: Emerg. Technol. 15, 380–391 (2007)
16. Hellinga, B.R., Fu, L.: Reducing bias in probe-based arterial link travel time estimates. Transp. Res. Part C: Emerg. Technol. 10, 257–273 (2002)
17. Bayen, M., Butler, J. and Anthony, D.: Mobile Millennium. Using Cell Phones as Mobile Traffic Sensors, UC Berkeley College of Engineering, CCIT, Caltrans, DOT, Nokia, NAVTEQ, pp. 1557–2269 (2008)
18. Tao, S., Manolopoulos, V., Rodriguez, S., Rusu, A.: Real-Time Urban Traffic State Estimation with A-GPS Mobile Phones as Probes. J. Transp. Technol. 2, 22–31 (2012)
19. Rose, G.: Mobile phones as traffic probes: practices, prospects and issues. Transp. Rev. 26, 275–291 (2006)
20. Chi, H. and Xavier, S.: A Fast Approach Towards Android Malware Detection. In: Computational Science and Its Applications—(ICCSA 2015). pp. 77–89. Springer, Berlin (2015)
21. Ken, H. Global Smartphone Sales Forecast for 88 Countries 2007 to 2017 (2013). https://www.strategyanalytics.com/strategy-analytics/blogs/devices/smartphones/smart-phones/2013/03/28/africa-smartphone-sales-will-jump-56-in-2013-#.VvMkWE_SnT9

22. Aydos, J. Hengst, B., Uther, W., Blair, A. and Zhang, J.: Stochastic Real-Time Urban Traffic State Estimation: Searching for the Most Likely Hypothesis with Limited and Heterogeneous Sensor Data, Ph.D. Thesis (2012)
23. Vlahogianni, E.I., Golias, J.C., Karlaftis, M.G.: Short-term traffic forecasting: Overview of objectives and method. Transp. Rev. **24**, 533–557 (2004)
24. Yin, H., Wong, S., Xu, J., Wong, C.: Urban traffic flow prediction using a fuzzy-neural approach. Transp. Res. Part C: Emerg. Technol. **10**, 85–98 (2002)
25. Alecsandru, C., Ishak, S.: Hybrid model-based and memory-based traffic prediction system. Transp. Res. Rec.: J. Transp. Res. Board **1879**, 59–70 (2004)
26. Zheng, W., Lee, D.H., Shi, Q.: Short-term freeway traffic flow prediction: Bayesian combined neural network approach. J. Transp. Eng. **132**, 114–121 (2006)
27. van Lint, J.W., Hoogendoorn, S., van Zuylen, H.J.: Freeway travel time prediction with state-space neural networks: Modeling state-space dynamics with recurrent neural networks. Transp. Res. Rec.: J. Transp. Res. Board **1811**, 30–39 (2002)
28. Anderson, J., Bell, M.: Travel time estimation in urban road networks. In: IEEE Conference on Intelligent Transportation System (ITSC'97), vol. 1997, 924–929 (1997)
29. Habtie, A.B., Ajith, A. and Dida. M.: Hybrid U-TDOA and A-GPS for VehiclePositioning and Tracking. Springer, Berlin, HAIS 2015, LNAI 9121, pp. 1–13, (2015)
30. Habtie, A.B., Ajith, A. and Dida. M.: Road traffic state estimation famework based on hybrid assisted global positioning system and uplink time difference Of arrival data collection methods. In: AFRICON, 2015. IEEE, p.308 (2015)
31. Habtie, A.B., Ajith, A. and Dida. M.: A neural network model for road traffic flow Estimation. NaBIC (2015) (in press)
32. Habtie, A.B., Ajith, A. and Dida. M.: Cellular network based real-time urban road traffic state estimation framework using neural network model estimation. In: IEEE SSCI (2015) (in press)
33. Gundlegard, D. and Karlsson, J.M.: Route classification in travel time estimation based on cellular network signaling. In: 12th International IEEE Conference on Intelligent Transportation Systems, 2009. (ITSC'09)
34. Minh, Q.T. and Kamioka, E.: Pinpoint: An efficient approach to traffic state estimation system using mobile probes. In: 2010 6th International Conference on Wireless Communications Networking and Mobile Computing (WiCOM), pp. 1–5 (2010)
35. Ferman, M.A., Blumenfeld, D.E. and Dai. X.: An Analytical Evaluation of a Real-Time Traffic Information System Using Probe Vehicles in Intelligent Transportation Systems. Taylor & Francis, London (2005)
36. Ferman, M.A., Blumenfeld, D.E. and Dai. X.: An analytical evaluation of a real-time traffic information system using probe vehicles. Intell. Transp. Syst. 23–34 (2005)
37. Manolopoulos, V., Tao, S., Rodriguez, S., Ismail, M., Rusu, A.: MobiTraS: a mobile application for a smart traffic system. In: 8th IEEE. International NEWCAS Conference (NEWCAS), vol. 2010, pp. 365–368 (2010)
38. Zhao, Q., Kong, Q.J., Xia, Y. and Liu, Y.: Sample size analysis of GPS probe vehicles for urban traffic state estimation. In: 2011 14th International IEEE Conference on Intelligent Transportation Systems (ITSC), pp. 272–276 (2011)
39. Zheng, F., Van Zuylen, H.: Urban link travel time estimation based on sparse probe vehicle data. Transp. Res. Part C: Emerg. Technol. **31**, 145–157 (2013)
40. Boubaker, S., Rehimi, F. and Kalboussi, A.: Comparative analysis of microscopic models of road traffic data. In: 2011 4th International Conference on Logistics (LOGISTIQUA), pp. 474–478 (2011)
41. Zhang, Y. and Liu, Y.: Comparison of Parametric and Nonparametric Techniques for Non-peak Traffic Forecasting. World Academy of Science, Engineering and Technology 2009
42. Wei, C., Lin, S., Li, Y.: Empirical Validation of Freeway Bus Travel Time Forecasting. Transp. Plan. J. **32**, 651–679 (2003)
43. Kisgyörg, L., Rilett, L.R.: Travel time prediction by advanced neural network. Civil Eng. **46**, 15–32 (2002)

44. Ishak, S., Alecsandru, C.: Optimizing traffic prediction performance of neural networks under various topological, input, and traffic condition settings. J. Transp. Eng. **130**, 452–465 (2004)
45. Habtie, A.B., Ajith, A. and Dida. M.: In-vehicle mobile phone-based road traffic flow estimation: a review. J. Netw. Innov. Comput. **2**, pp. 331–358 (2013)
46. Topuz. V.: Hourly Traffic Flow Predictions by Different ANN Models, pp. 1–18 (2010)
47. Shan, Z., Wang, Y. and Zhu, Q.: Feasibility study of urban road traffic state estimation based on taxi GPS data. In: 2014 IEEE 17th International Conference on Intelligent Transportation Systems (ITSC), pp. 2188–2193 (2014)
48. Guan, W., Deng, Z., Ge, Y., Zou, D.: A practical TDOA positioning method for CDMA2000 mobile network. In IEEE International Conference on Wireless Communications, Networking and Information Security (WCNIS), vol. 2010, pp. 126–129 (2010)
49. Liu, K., Yamamoto, T., Morikawa, T.: Feasibility of using taxi dispatch system as probes for collecting traffic information. J. Intell. Transp. Syst. **13**, 16–27 (2009)
50. Chen, Y., Gao, L., Li, Z., p. and Liu, Y.C.: A new method for urban traffic state estimation based on vehicle tracking algorithm. In: Intelligent Transportation Systems Conference (ITSC 2007), vol. 2007, pp. 1097–1101. IEEE (2007)
51. Quiroga, C.A.: An integrated GPS-GIS methodology for performing travel time studies. No. 98–08771 (1997)
52. Kong, Q.J., Li, Z., Chen, Y., Liu, Y.: An approach to urban traffic state estimation by fusing multisource information. Intell. Transp. Syst. **10**, 499–511 (2009)
53. Qiankun, Z., Qingjie, K., Yingjie, X. and Yuncai, L.: An improved method for estimating urban traffic state via probe vehicle tracking. In: 2011 30th Chinese Control Conference (CCC), pp. 5586–5590 (2011)
54. Kong, Q.J., Chen, Y., Liu, Y.: A fusion-based system for road-network traffic state surveillance: a case study of Shanghai. Intell. Transp. Syst. Mag. **1**, 37–42 (2009)
55. Behrisch, M., Bieker, L., Erdmann, J., Krajzewicz, D.: SUMO-Simulation of Urban MObility-an Overview. In: The Third International Conference on Advances in System Simulation (SIMUL), vol. 2011, pp. 55–60 (2011)
56. SUMO: Simulation of Urban Mobility. http://www.dlr.de/ts/desktopdefault.aspx/tabid-9901/
57. JOSM: Java OSM Editor. https://josm.openstreetmap.de
58. Adusei, I.K., Kyamakya, K. and Jobmann, K.: Mobile positioning technologies in cellular networks: an evaluation of their performance metrics. In: Proceedings of the MILCOM 2002, pp. 1239–1244 (2002)
59. Scorer, A.: Vehicle Location and Navigation Systems. By Yilin Zhao. Artech House, 1997. 345 pages,£ 65 Hardback. ISBN: 0-89006-861-5. J. Navig. **51**, 445–447 (1998)
60. Hammerstrom, D.: Neural networks at work. IEEE Spectr. **30**, 26–32 (1993)
61. Adeoti, O.A., Osanaiye, P.A.: Performance analysis of ANN on dataset allocations for pattern recognition of bivariate process. Math. Theory Model. **2**, 53–63 (2012)
62. Ranganathan, A.: The Levenberg-Marquardt algorithm. Tutor. LM Algorithm **11.1**, 101–110 (2004)

Attack Detection Using Evolutionary Computation

Martin Stehlik, Vashek Matyas and Andriy Stetsko

Abstract Wireless sensor networks (WSNs) are often deployed in open and potentially hostile environments. An attacker can easily capture the sensor nodes or replace them with malicious devices that actively manipulate the communication. Several intrusion detection systems (IDSs) have been proposed to detect different kinds of active attacks by sensor nodes themselves. However, the optimization of the IDSs w.r.t. the accuracy and also sensor nodes' resource consumption is often left unresolved. We use multi-objective evolutionary algorithms to optimize the IDS with respect to three objectives for each specific WSN application and environment. The optimization on two detection techniques aimed at a selective forwarding attack and a delay attack is evaluated. Moreover, we discuss various attacker strategies ranging from an attacker behavior to a deployment of the malicious sensor nodes in the WSN. The robustness of the IDS settings optimized for six different attacker strategies is evaluated.

Keywords Attacker strategy · Evolutionary algorithm · Intrusion detection system · Multiobjective optimization · Wireless sensor network

1 Introduction

Wireless Sensor Networks (WSNs) are highly distributed networks often deployed in open or even hostile environments. Sensor nodes are usually small low-cost and resource constrained devices with no tamper resistance employed and can be easily captured by an attacker. Furthermore, malicious devices considered as benign by other sensor nodes can be deployed in the network. Possible attacks on WSNs range from passive eavesdropping where an attacker listens on the ongoing traffic in promiscuous mode to active interfering and manipulating of the communication.

M. Stehlik (✉) · V. Matyas · A. Stetsko
Masaryk University, Brno, Czech Republic
e-mail: xstehl2@fi.muni.cz

© Springer International Publishing AG 2017

A. Abraham et al. (eds.), *Computational Intelligence in Wireless Sensor Networks*,
Studies in Computational Intelligence SCI 676, DOI 10.1007/978-3-319-47715-2_5

99

In this chapter, we aim at the detection of such active attacks by sensor nodes themselves with respect to restricted capabilities of the sensor nodes. Each sensor node can be equipped with an intrusion detection system (IDS) [1]. Thus, an entire network area can be monitored for malicious behavior in a distributed manner. Several detection techniques have been proposed to detect various kinds of attacks on WSNs. Unfortunately, many of them are proposed for a specific case or their optimization is left unresolved. Moreover, incorporating an IDS brings necessary additional demands on sensor nodes' resources. In our work, we aim at automatic configuration of these detection techniques for a specific application scenario, topology and environment, using evolutionary computation having both the IDS accuracy and resource consumption in mind.

In this work, we demonstrate the functionality of our optimization framework consisting of a simulator and an optimization engine utilizing multi-objective evolutionary algorithms on two detection techniques proposed in [2]. The detection techniques are aimed at two kinds of active attacks—the selective forwarding attack and the delay attack. Our optimization framework provides Pareto front approximations consisting of different IDS settings with respect to three objectives—false positives, false negatives and memory consumption. We elaborate on these IDS settings found by evolution. Furthermore, we discuss various attacker strategies that can be used by an attacker and discuss the robustness of the IDS settings found for a specific attacker strategy in cases where another attacker strategy is used.

The contributions of this work are the following:

1. We provide a complex optimization framework for optimization of IDSs for WSNs. Various multi-objective evolutionary algorithms (NSGA-II, SPEA2, IBEA) can be used for optimization. The optimization framework can be easily extended to solve another optimization issue in WSNs.
2. We demonstrate the functionality of the optimization framework through extensive experiments on detection techniques detecting two different attacks— selective forwarding and delay attack.
3. We discuss and evaluate various attacker strategies. The IDS is optimized for six different attacker scenarios. Robustness of optimized IDS settings for each of the attacker strategy is evaluated on every other attacker strategy.

The chapter is organized as follows. In Sect. 3, we present our intrusion detection system and describe detection techniques used to detect the selective forwarding attack and the delay attack. Our optimization framework utilizing evolutionary algorithms and the metrics we use to compare resulting populations are described in Sect. 4. Various attacker strategies that can be used by an attacker are discussed in Sect. 5. In Sect. 6, we specify the settings of the evaluated IDS and of the experiments. Also, our test case WSN is presented. Experiment results are elaborated upon in Sect. 7. Related work is discussed in Sect. 2 and the chapter is concluded in Sect. 8.

2 Related Work

The related work is divided into two parts. First, we discuss related detection techniques for WSNs. Second, we discuss computational intelligence-based solutions for IDSs in WSNs.

2.1 Selective Forwarding and Delay Attacks

Selective forwarding attack has been among the most discussed attacks in WSNs during recent period. Karlof and Wagner [3] introduced selective forwarding attack in WSNs and discussed the possibilities of an attacker to place a malicious sensor node on a path between data source and base station. da Silva et al. [1] defined a "retransmission rule" as listening to a packet by an IDS whether it was forwarded by monitored sensor node or not. Krontiris et al. [4] set a threshold value for the percentage of packets dropped to 20 %. Another proposals of detection techniques for selective forwarding attack where the parameter setting is left unresolved can be found, e.g., in [5, 6]. To the best of our knowledge, no work has been published on such a complex parameter optimization for collaborative detection of selective forwarding attack.

The delay attack detection has not been discussed very much. da Silva et al. [1] defined a "delay" rule as a timeout before which a retransmission by monitored sensor node has to occur. Liu et al. [7] used the forwarding delay time measurement for their complex insider detection technique but its parametrization was left unresolved. To the best of our knowledge, we are the first who present a complex collaborative detection technique aimed at the delay attack detection.

2.2 Computational Intelligence-Based Solutions

A few papers utilize computational intelligence-based solutions to secure WSNs with IDSs.

Khanna et al. used single-objective evolutionary algorithms for several optimization issues in WSNs [8–10]. In [8], the authors aimed at minimizing power consumption while maximizing the coverage and exposure by switching the sensor nodes to the following states: (1) *inactive sensor node*; (2) *active sensor node*; (3) *cluster-head*; and (4) *inter-cluster router*. In [9], Khanna et al. presented an approach on a WSN deployment how it can be optimized dynamically and considered from the security point of view as well. The authors considered deployment of cluster-heads and inter-cluster routers allowing encryption and authentication, respectively. Finally, in [10], Khanna et al. incorporated *local monitoring nodes* into the network that observed suspicious behavior like data message patterns, message collisions and sensor

positioning in their neighborhood. An IDS placement problem was addressed by
adding evaluation of *local monitoring nodes* to the fitness function used in [9]. In all
papers [8–10], all the objectives were blended into a single fitness function.

To the best of our knowledge, Heady et al. were the first introducing
evolutionary algorithms to the area of IDSs for wired networks and the work of
Khanna et al. [10] is the only work on optimization of IDSs [11] using evolutionary
algorithms for WSNs except of our IDS optimization framework. We introduced
multi-objective evolutionary algorithms to IDS in WSNs in [12].

Several works on another metaheuristics utilized for IDSs in WSNs can be found.
Banerjee et al. [13, 14] used swarm intelligence—ant colony—for localization of
the source of intrusion. The ants traverse the sensor nodes via edges that connect
the neighboring sensor nodes. Adjacent sensor nodes with maximum number of
violations represented as pheromone are preferred. When an ant visits an edge, the
application of the local update rule makes the edge pheromone level diminish to
the edge becoming less attractive. Mukherjee and Sen [15] detect intentionally sent
erroneous data on the base station using neural networks. A hierarchical network is
assumed where non-leaf sensor nodes aggregate data from their descendants. The
neural networks predict the sensed data of a node N, provided the data reported by
neighbors of the node N are given.

3 Intrusion Detection

Sensor nodes are vulnerable devices given by the nature of wireless communication
and because of other limitations. They are usually deployed in open or even hostile
environments where they can be easily manipulated or even stolen by an attacker. In
order to keep their price low and to consume little energy, the nodes consist usually of
a simple hardware where some conventional security countermeasures are unusable.
Furthermore, a typical sensor node is not tamper resistant.

Attackers on WSNs can be categorized into two following classes [3, 16]:

1. *Passive attacker* uses his own device to listen on the ongoing communication to
   obtain sensitive data without any manipulation of the traffic within the WSN.
2. *Active attacker* uses his own device to disrupt the functionality of the WSN,
   manipulates the content of the ongoing packets, drops the packets, presents a
   fake identity to other sensor nodes or jams legitimate transmissions.

While passive attacks can be prevented by encryption, intrusion detection systems
(IDSs) are used to detect some kinds of active attacks. In this chapter, we focus on
distributed IDSs [1] where sensor nodes themselves monitor the overall network area
by promiscuous listening on the transmissions among their neighbors.

We demonstrate the usability and benefits of IDS optimization using evolutionary
computation for WSNs to detect *selective forwarding* and *delay* attacks. The description of these attacks and of our detection techniques proposed to detect these attacks
follows.

3.1 Selective Forwarding Attack

Selective forwarding is one of the most widely discussed attack in WSNs [1, 3–7, 17]. When performing a selective forwarding attack, an attacker inserts a malicious sensor node into a WSN, where this node is believed as legitimate by the other (benign) sensor nodes. Once becoming a participant in packet routing, such a malicious sensor node can easily drop all the packets routed via itself (*blackhole* attack [3]) or can forward them selectively based on their contents, sensor measurements, source IDs or just randomly. Various attacker strategies are discussed in Sect. 5.

3.2 Delay Attack

The prerequisites for the delay attack are similar as for the selective forwarding attack—a malicious sensor node has to become a member of a routing tree in a WSN. Consequently, instead of dropping, the packets are intentionally delayed but finally forwarded. This kind of attack is aimed at time sensitive applications for WSNs where the delivery time of the packets to the BS is of a crucial importance. Such applications involve fire detection, people or animal movement detection and others. The same attacker strategies as for the selective forwarding attack can be applied for the delay attack.

3.3 Detection Techniques

We use *distributed* detection [1] where the IDS runs on each sensor node deployed in a WSN. Thus, the entire network area can be monitored to detect malicious behavior by sensor nodes themselves. However, this approach requires additional resources (e.g., memory and energy).

We provide two approaches of distributed detection:

1. *Non-collaborative detection*—no additional communication among the IDS nodes is required.
2. *Collaborative detection*—IDS nodes collaborate on the decision about monitored sensor nodes using exchange of *voting* packets.

We use the following notations to explain the functionality of the IDS [18]:

Notation 1 *The set $A = \{a_1, \ldots, a_{n_m}\}$ is a set of all malicious nodes in a network.*

Notation 2 *The set $C = \{c_1, \ldots, c_{n_b}\}$ is a set of all benign nodes in a network.*

Notation 3 *The function $x : \mathbb{N} \to \mathbb{N}$ takes a sensor node index as an argument, and returns a number of the neighbors that consider this node benign.*

Notation 4 *The function* $y : \mathbb{N} \to \mathbb{N}$ *takes a sensor node index as an argument, and returns a number of the neighbors that consider this node malicious.*

Notation 5 *The function* $n : \mathbb{N} \to \mathbb{N}$ *takes a sensor node index as an argument, and returns a number of the neighbors of this node.*

Notation 6 *The function* $m : \mathbb{N} \to \mathbb{N}$ *takes a sensor node index as an argument, and returns the amount of memory (in bytes) used by an IDS on this node.*

Neighbor $b_k \in C \cup A$ of a node $c_j \in C$ is each node such that c_j overheard at least one packet from b_k since the beginning of the WSN operation time.

Monitored neighbor $b_l \in C \cup A$ of a node $c_j \in C$ is such a *neighbor* of the node c_j that the IDS running on the node c_j collects the statistics of the packet forwarding of the node b_l. The selection process for the set of the monitored neighbors is described in Sect. 3.4.

Solution s is a specific configuration of the IDS in a form of a detection technique and specific values given to each of the parameters used by that technique.

Ranges of values of IDS parameters discussed in the following text are then discussed in more detail in Sect. 6.3.

3.4 Non-collaborative Detection of Selective Forwarding Attack

In [12], we evaluated multi-objective evolutionary algorithms (MOEAs) on a simple IDS detecting selective forwarding attack. An IDS was running on each sensor node and continuously monitoring its own sent and also overheard packets addressed to all monitored sensor nodes whether they were forwarded or dropped by those monitored sensor nodes. Since the collaborative version used for the experiments is an extension of the non-collaborative version, we first describe the functionality of the non-collaborative version.

The basic principle is illustrated in Fig. 1. The black dots represent sensor nodes that are placed within communication range of sensor node $b_i \in C \cup A$ and, thus, can monitor b_i for selective forwarding attack. However, the number of monitored neighbors is limited to p_1 (*max monitored nodes*), not only due to memory reasons— the IDS can have incomplete information about furthest neighbors (the IDS nodes can be interfered, far from the monitored node or hidden behind an obstacle) causing additional false positives. Thus, each IDS monitors at most p_1 nearest neighbors (according to received signal strengths). The arrows represent routing directions of the packets—b_i forwards all received packets to a parent node $b_j \in C \cup A$. The IDS maintains a table, where each of p_1 rows corresponds to a certain monitored node. The table contains the number of packets received (PR) and forwarded (PF) by each monitored node.

The IDS stores all overheard packets addressed to all monitored neighbors in a single buffer limited to p_2 packets (*buffer size*). Each time a packet P addressed to

Fig. 1 Non-collaborative
intrusion detection

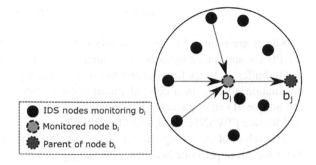

a monitored node b_i is overheard by the IDS, the PR counter of b_i is incremented
and packet P stored in the buffer. Once the node b_i forwards the packet P, the IDS
increments PF of the node b_i and packet P is removed from the buffer. In case the
packet P is being the oldest one and the buffer is full, it is removed from the buffer
without incrementing the PF counter.

Finally, during the evaluation phase, a sensor node b_i is considered as attacker by
the IDS node if two following conditions hold:

1. The IDS node has overheard (or sent) at least p_3 packets (*min received packets*)
   addressed to b_i.
2. The ratio of forwarded and received packets (*PF/PR*) is lower than p_4 (*detection
   threshold*).

Objective function 1 The number of *false negatives* (fn) of a solution s is calculated
as follows:

$$fn(s) = \frac{1}{|A|} * \sum_{a_i \in A} \frac{x(a_i)}{n(a_i)}. \tag{1}$$

The values of fn range from 0 to 1. If every malicious node in the network is
correctly detected by all its neighbors, fn is equal to 0 and if none of malicious
nodes is detected by any of its neighbors, fn equals to 1.

Objective function 2 The number of *false positives* (fp) of a solution s is calculated
as follows:

$$fp(s) = \frac{1}{|C|} * \sum_{c_i \in C} \frac{y(c_i)}{n(c_i)}. \tag{2}$$

The values of fp range from 0 to 1. If every benign node in the network is
considered benign by all its neighbors, fp is equal to 0 and if all benign nodes are
considered malicious by all its neighbors, fp equals to 1.

Objective function 3 The consumed *memory* (*mem*) in a solution s is calculated as
follows:

$$mem(s) = 8 * p_1 + 16 * p_2, \tag{3}$$

8 bytes are required for every monitored neighbor (4 bytes for node ID, 2 bytes for PR counter and 2 bytes for PF counter) and 16 bytes are required for one slot in the buffer (4 bytes for source address, 4 bytes for receiver address, 4 bytes for destination address in a case of multiple base stations in the WSN and 4 bytes for unique ID of a packet). The memory demands come from our real security middleware ("WSNProtectLayer" for WSN [19]).

The values of *mem* range from 0, where the IDS is potentially switched off, to 288 bytes for our upper bounds of $p_1 = 30$ and $p_2 = 3$ used for the selective forwarding attack. For the delay attack, the upper bounds of $p_1 = 30$ and $p_2 = 10$ results in the maximum memory consumption of 400 bytes. The upper bound of p_1 (*max monitored nodes*) is based on our experiments. There is no significant improvement of any of the objectives with p_1 higher than 30. See Sect. 7 for more details. The upper bounds of p_2 (*buffer size*) are based on throughput analysis in the simulator.

3.5 Collaborative Detection of Selective Forwarding Attack

Collaborative detection of the selective forwarding attack that we first presented in [2] is an extended version of the non-collaborative detection discussed in Sect. 3.4. The idea behind the collaborative approach basically comes from [4] regarding to voting scheme and time windows. However, we enriched the collaborative approach presented in [4] by parameters "voting threshold" and "minimum received votes".

The monitored nodes are not evaluated by the IDS nodes at the end of the simulation. Instead, the simulation time is divided into windows of size p_5 (*time window*). The time windows are of the same fixed size among all the IDS nodes, but they are asynchronous—the first window of each IDS node is started randomly within the time interval of p_5. At the end of each time window, all monitored neighbors are evaluated by the IDS node and if an attack was detected, a voting process can be executed.

An example situation is depicted in Fig. 2. IDS node c_k monitors (among others) node b_i and detected too many packets dropped by b_i in a time window marked as "Attack!" in Fig. 3. This decision is based on the same principle as for the non-collaborative IDS discussed in Sect. 3.4. Since this time, IDS node c_k considers a node b_i malicious "locally"—still no alert is produced.

Then, the decision of the node c_k is checked with all neighbors of c_k. Thus, c_k broadcasts a voting request to its neighbors (arrows from c_k in Fig. 2 point the neighbors of c_k that can also monitor node b_i). Each of the asked nodes that also monitors node b_i answers at the end of its own time window. If an asked IDS node consider b_i an attacker (either just locally or globally), it answers positively, otherwise negatively. Node c_k waits the following time window to collect the responses. Finally, the monitored node b_i is considered an attacker "globally" by c_k and c_k produces an alert if two following conditions hold:

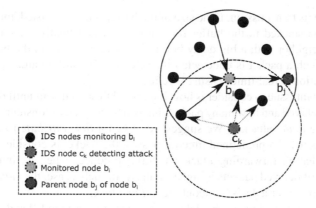

Fig. 2 Collaborative intrusion detection

Node c_k monitoring node b_i:

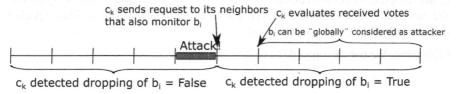

Fig. 3 Time windows. Since the end of the window in which the attack was detected, node c_k always votes positively about node b_i. Node b_i can be globally considered malicious by c_k since the end of window following the one in which the attack was detected if the voting result is positive

(i) At least p_6 votes (*min votes received*) were received.
(ii) The ratio of positive and all responses is at least p_7 (*voting threshold*).

Malicious nodes can falsely report to the IDS nodes to defend another malicious node. We consider this to be an specific attacker strategy that is discussed in Sect. 5.

Objectives We use the same three objectives as in Sect. 3.4.

3.6 Collaborative Detection of Delay Attack

Time related attacks result in long delays and traffic imbalance [20]. We believe that a WSN should guarantee the delivery time for some applications (e.g., movement monitoring or fire detection). In [1], the *delay detection rule* is defined as follows: "The transmission of a message by a monitor's neighbor must occur before a *defined timeout*". We adapt this rule to our IDS detecting delay attack.

A technique that is similar to that one used for the selective forwarding attack can be extended to detect intentional delays. Using the buffering technique discussed above, a packet can finally be considered forwarded even though some malicious node delayed its transmission. If the size of the buffer is not exhausted, the monitored packet can be stacked in the buffer for a long time and finally forwarded by the monitored neighbor with a big delay before being removed from the buffer by the IDS. Thus, such a packet is undetected for selective forwarding attack, yet useless for a base station if real-time sensing is required.

An important issue to consider is how long the IDS should wait until the packet is considered delayed and how many packets have to be delayed to consider a monitored neighbor as a *delay* attacker. We suggest to assign a *time attribute* to each of the buffered packets. If a predefined timeout passes, the packet is considered delayed. As for the selective forwarding attack, an alert is produced when p_5 time units pass and the ratio of delayed packets is higher than p_4. In such a case, the *majority voting scheme* is applied for a decision about the *delay attack*.

Our detection technique of the delay attack was proposed in [2] and is evaluated thoroughly in this chapter. The detection technique extends the selective forwarding attack detection specified in Sects. 3.4 and 3.5. As we focus on collaborative detection in this chapter, we do not consider non-collaborative detection of the delay attack. We incorporate another parameter p_8 (*delay timeout*) that is a timeout when a packet in the IDS buffer is marked as delayed.

Objectives We use the same three objectives as in Sect. 3.4.

4 Computational Intelligence-Based Optimization

In this section, we present our optimization framework that was designed in [18]. The optimization framework consists of two main components: a *network simulator* and an *optimization engine*. The simulator is used for evaluation of the candidate solutions (in a form of IDS configurations) that were designed by the optimization engine. Based on the simulation results in a form of metrics described in Sect. 3.4 (false positives, false negatives, memory), the optimization engine produces a new generation of solutions.

We enhance the optimization framework to easily distribute the computation of the individuals in each generation to multiple computers using BOINC (Berkeley Open Infrastructure for Network Computing) distributed computing platform [21] and to use MOEAs. The efficiency of MOEAs for the problem of IDS optimization for the WSNs was evaluated in [12]. Also the MOEAs configuration issue was addressed there. The findings are utilized in this work.

4.1 Simulator

We use the MiXiM simulator [22] that is based on OMNeT++ simulation platform [23]. The selection of the simulator is based on our previous results on the comparison of various simulators with the reality [24]. MiXiM is a discrete event simulator with a good support of wireless channel modeling and support of all communication layers of the current sensor nodes.

However, as we showed in a previous work [24], a proper calibration of all the models is required to obtain sufficiently accurate results. The wireless channel model is based on a *log normal shadowing* [25] that is widely used for wireless communication modeling. Two parameters of the model has to be set up according to modeled environment. The values of the parameters can be either based on measurements of wireless signal propagation in the target area or based on recommended values for target environment type [25]. For experiments in this chapter, we use recommended values for an outdoor environment (path loss exponent equals to 2 and standard deviation of the attenuation equals to 2). All simulated sensor nodes transmit packets with power equals to −25 dBm (transmission level 3 of TelosB [26] sensor nodes). See [24] for more information about our calibration approach or [25] for the theoretical background.

4.2 Evolutionary Algorithms

We found evolutionary algorithms to be a very efficient metaheuristic solving optimization of the IDS for WSNs. As we showed in [12], optimal or near-optimal results can be found in a feasible time. For a problem solved in this chapter, we are not able to compute exhaustive search even with the computational grid we have at our disposal. Thus, we compare the results found by evolution with much more time demanding sampling.

In our early work [18], we used *single-objective* evolutionary algorithms, where a fitness function blending all three objectives with user-specified weights had to be defined. The main disadvantage was the amount of required experience in weight definition. If the network operator wanted to change the weights, the optimization process had to run again. Thus, based on [12], we recommend to use *multi-objective* evolutionary algorithms (MOEAs) that eliminate the mentioned problem.

Using MOEAs, the network operator can choose any IDS setting from the *Pareto front*[1] [27] approximation and change the selection to another optimized one any time according to current requirements. Based on results from [12], where we evaluated 48 different MOEA's configurations for two widely used algorithms—NSGA-II [28] and SPEA2 [29], we configure the evolution as follows in this chapter: The population consists of 200 individuals, 200 generations are computed, probability of multi-point crossover is 0.5 for all experiments. Mutation probability of each parameter is 0.01 or 0.25 for results marked as "Evo #1" and "Evo #2", respectively. If mutation is

---

[1]Pareto front is a set of non-dominated solutions with respect to all objectives. Thus, a network operator can easily choose between a solution A with a better IDS accuracy but higher resource consumption or solution B with a worse IDS accuracy but lower resource consumption. Solution C, that is dominated by A and B in all objectives is dominated and, thus, is not a member of the Pareto front.

performed, the value is shifted randomly within the interval of 10 % of the overall parameter range. NSGA-II is used in all experiments.

Performance Metrics Since we are not able to compare the Pareto front approximations found either by sampling or by evolution with the true Pareto front, we use two metrics to mutually compare the different optimization strategies. These metrics are also used to compare the IDS performances across different attacker strategies.

1. *Hyper-volume indicator* [27, 30]—Hyper-volume indicator is given by the calculation of a volume of the objective space that is dominated by evaluated Pareto front approximation. A reference point R that is dominated by each solution in the Pareto front approximation has to be established—as an upper bound of each of the objectives. The reference point serves as an upper bound of the dominated objective space for the volume calculation. Minimal values of the coordinates of the point R are maximal values of each of the objectives across the population of the evaluated Pareto front approximation.

   In all our experiments where the hyper-volume indicator is calculated, we normalize the objective function *memory* dividing its output by its upper bound (288 for the selective forwarding attack and 400 for the delay attack). The results of the normalized *memory* function range from 0 to 1. The value of the reference point is established to: $R = [1, 1, 1]$ for all calculations of the hyper-volume indicator in this work to enable mutual comparisons.

   The value of the hyper-volume indicator ranges between 0 (potential unrealistic worst single solution of the IDS setting in the Pareto front approximation consuming all possible memory in spite of evincing the values of false positives and false negatives equal to 1) and 1 (potential unrealistic ideal single solution of the IDS setting in the Pareto front approximation consuming no memory and evincing the values of false positives and false negatives equal to 0).

   We use the software presented in [31] to calculate the hypervolume-indicator.

2. *Coverage metric* [32]—Coverage metric enables mutual comparison of two different Pareto front approximations. It is used to calculate the percentage of solutions found in the Pareto front approximation A that are not dominated by any solution found in the Pareto front approximation B, and vice versa.

   The value of the coverage metric when calculating percentage of solutions in A dominated by solutions in B ranges from 0 % when none of the solutions in A is dominated by any solution in B to 100 % when each solution in A is dominated by at least one solution in B. Note that if a mutual comparison is required, we also have to calculate the percentage of solutions in B dominated by solutions in A resulting in two numbers all together as the coverage metric output.

3. *Number of simulations*—Since the evaluation of each individual by simulator is time demanding (approx. 5–8 min), we also provide a number of simulation runs needed within the whole optimization process.

4. *Number of non-dominated solutions*—This metric denotes the number of non-dominated solutions in the resulting Pareto front approximation.

5   Attacker Strategies

The aim of this section is to elaborate on the strategies an attacker can take to efficiently selectively forward or drop the packets. We also discuss the impact of each of the strategy on the IDS performance. Furthermore, we provide a set of recommendations with respect to corresponding attacker strategies for a network operator before setting up the IDS optimization process.

We divide the concepts of attacker strategies into the three following categories:

1. *Attacker behavior*—content-based selection, ratio of dropped and all received packets by a malicious node, malicious voting.
2. *Deployment of malicious nodes*—strategy of malicious sensor nodes deployment in a WSN.
3. *Number of deployed malicious nodes*—ratio of malicious and all sensor nodes in a WSN.

Note that if only the selective forwarding attack is discussed in the following text, the attacker strategy for the delay attack would be equivalent. The only difference would be that selected packets on malicious nodes are delayed instead of being dropped.

5.1   Attacker Behavior

A malicious sensor node can take various strategies w.r.t. the decision whether a packet should be forwarded or not. In the ultimate case, all the packets can be dropped by a malicious node $a_i \in A$, effectively changing the selective forwarding attack to a *blackhole* attack [3]. However, the blackhole attack can be easily detected either by our distributed IDS running on the neighboring sensor nodes (dropping ratio higher than any value of the *detection threshold* p_4 is detected). Furthermore, the base station not receiving any packets from such sensor node a_i and also from all the descendants of the a_i can easily detect the attack.

We discuss different approaches of the selection of the packets to be dropped that an attacker can use to decrease the probability of being detected. Furthermore, we discuss impacts of malicious voting.

Dropping Ratio Random dropping with a given ratio is probably the most simple way of selective forwarding attack. The attacker specifies the percentage s_1 of dropped packets by a malicious node $a_i \in A$ as the only parameter for the random dropping. Consequently, the packets that should be forwarded by a_i are dropped with the probability s_1.

Recommendation for the IDS We recommend usage of the collaborative version of the IDS optimized for a target WSN as discussed in Sect. 3. A network operator decides carefully on the minimum percentage s_1 of the packets dropped by the sensor nodes that is considered unacceptable. Such a dropping rate is to be set up for all the simulated malicious sensor nodes. The optimization process adapts all the IDS parameters, particularly the parameter p_4 (*detection threshold*), taking into account also packet losses caused by interference, congestion and other aspects of the unreliable wireless communication. When such an optimized IDS is used in a real WSN and any malicious (or malfunctioning) sensor node drops packets with even a higher frequency, then such sensor node is detected even more reliably. In the other case, if the dropping ratio of a monitored node is lower, the selective forwarding attack may not be detected because such a behavior is considered acceptable.

Dropping in Bulk Another behavior of a malicious node $a_i \in A$ can be dropping the packets in a bulk. It means that all or some percentage s_1 of the packets can be dropped by a_i, but only within specific time intervals with a minimum length of s_2 seconds for each of them. This strategy can be used to suppress transmitted packets during some specific sensitive period (e.g., dropping packets while a physical attacker is approaching the area of a WSN that detects people movement and protects some environment against unauthorized access).

Recommendation for the IDS We recommend to set up the parameter p_5 (*time window*) of the IDS to the length of $\frac{s_2}{2}$ at maximum for the minimum length of the interval s_2 seconds that is considered unacceptable. This countermeasure ensures that all the IDSs in the neighborhood of the malicious sensor node a_i have a chance to detect dropping (if it is higher than p_4—*detection threshold*) within a single time window—not decreased by a non-attacking phase of the malicious node a_i. If such an upper bound for the parameter p_5 is set up during the optimization, the other IDS parameters are optimized accordingly—particularly the parameter p_3 (*minimum received packets*) depends on the length of the time window. Note that both recommendations for the *dropping ratio* and the *dropping in bulk* complement each other.

Content Based Dropping A malicious node $a_i \in A$ can drop the packets based on their contents distinguishing important and non-important packets. To give an example of such a situation, the important packets can inform the base station about the presence of an intruder while the non-important packets can be periodically sent "still-alive" packets informing the base station that the sensor nodes are in an operation mode.

Recommendation for the IDS We recommend to extend the IDS by a separate buffer and table for such important packets—a monitoring node will separately monitor important and non important packets. This would mean additional requirements on the amount of memory consumed by the IDS (approx. doubled). However, the optimization process is analogous for both types of the packets. The definition of the malicious behavior in the context of the *dropping ratio* and the *dropping in bulk* discussed above is left to the network operator for both types of the packets.

Source Based Dropping A malicious node $a_i \in A$ can drop the packets based on their source addresses in order to drop packets from, e.g., some specific location.

Recommendation for the IDS In some cases, we believe that such a selective dropping based on the source address can be detected without any changes of the IDS functionality. However, the network operator should consider the impact of the selective forwarding based on the source address on the overall percentage of dropped packets by the monitored sensor node. Maintaining separated IDS buffers and tables for each source address can increase the sensitivity of the IDS on such source based dropping at the cost of increased memory consumption.

Delay Interval When a delay attack is executed, a malicious node $a_i \in A$ can delay the packets (by any strategy discussed above) for a fixed timeout d_1 seconds or randomly within interval $\langle d_2, d_3 \rangle$ seconds.

Recommendation for the IDS A network operator should decide carefully on the minimum timeout d_1 of packets delayed by the sensor nodes that is considered unacceptable. Such a delay timeout is to be set up for all the simulated malicious sensor nodes. The optimization process adapts all the IDS parameters, particularly the parameter p_4 (*detection threshold*) and the parameter p_8 (*delay timeout*), taking into account also packet losses caused by interference, congestion and other aspects of unreliable wireless communication. When such an optimized IDS is used in a real WSN and any malicious (or malfunctioning) sensor node delays packets with even a higher frequency or a longer timeout, then such sensor node is detected even more reliably. In the other case, if the dropping ratio of a monitored node is lower or the delay timeout is shorter, the delay attack may not be detected because such a behavior is considered acceptable.

Malicious Voting A malicious node(s) can falsely vote for benignity in order to defend another malicious node.

Recommendation for the IDS Considering such attack on the IDS, a network operator should adjust p_7 (*voting threshold*). For example, if it is assumed that up to 20 % of neighbors can be malicious, the *voting threshold* should be lower than 0.8 because up to 20 % neighbors vote falsely for benignity. The precise value of the *voting threshold* is subject of the optimization and consequently preferences—there is a trade-off between false positives and false negatives.

5.2 Deployment of Malicious Nodes

An attacker can deploy malicious sensor nodes into a WSN in specific "patterns" [33, 34]. However, in most papers, the deployment strategy of the sensor nodes into a WSN is left unresolved [5, 17] or the placement of malicious sensor nodes is selected randomly [4, 6, 7].

Fig. 4 Topology of the
evaluated WSN for the
random attacker strategy.
The sensor nodes are
represented by *circles* while
the base station is
represented by the *red
diamond*. The *black circles*
represent malicious sensor
nodes for the scenario with
2 % malicious sensor nodes
and together with the *gray
circles* for the scenario with
10 % malicious sensor nodes
(color figure online)

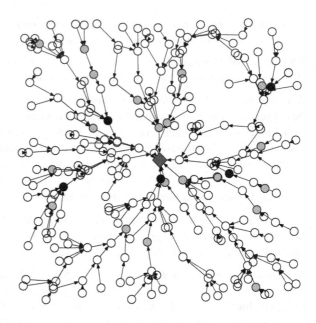

Influence of several realistic malicious nodes deployment strategies on the IDS
performance is discussed in this subsection. It might be impossible for the network
operator to predict the attacker deployment strategy. Thus, the impact of all discussed
attacker strategies on the IDS parametrization is evaluated in Sect. 7 to give a clue
how the malicious sensor nodes can be deployed in the simulations to obtain as robust
IDS parameters as possible.

Random Attacker Strategy The random attacker strategy is the most widely considered strategy by IDSs for WSNs [4, 6, 7]. In this approach, the malicious sensor
nodes are inserted into the WSN on random positions. However, we assume this
attacker strategy being far from the behavior of a real attacker in most cases. An
attacker can have access to or may be interested in only a specific part of the environment. However, this attacker strategy can be utilized for IDS optimization. If a
sufficient number of sensor nodes is considered, the random deployment can cover
different places (close or far from the base station, with sparsely or densely deployed
sensor nodes, etc.) at the same time.

We parameterize this attacker strategy by a percentage of nodes controlled by an
attacker by s_3. In Fig. 4, we give an illustration of the random attacker strategy that
we evaluate in Sect. 7.

Center Drop Attacker Strategy The goal of an attacker within this strategy is to
compromise sensor nodes surrounding the base station. Alternatively, an attacker can
choose an arbitrary target place in the WSN and compromise its surrounding sensor
nodes.

Fig. 5 Topology of the evaluated WSN for the center drop attacker strategy. The sensor nodes are represented by *circles* while the base station is represented by the *red diamond*. The *black circles* represent malicious sensor nodes for the scenario with 2 % malicious sensor nodes and together with the *gray circles* for the scenario with 10 % malicious sensor nodes (color figure online)

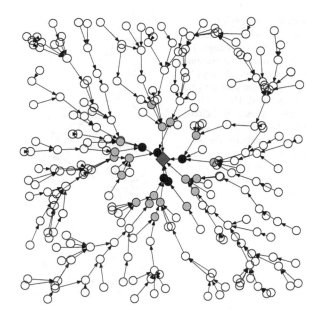

In this chapter, we evaluate a situation where the sensor nodes surrounding the base station are malicious. We parameterize this attacker strategy by s_4—percentage of the malicious sensor nodes in the WSN ordered by a distance from the base station. In Fig. 5, we give an illustration of the random topology that we evaluate in Sect. 7.

Direct Center Attacker Strategy In the direct centre attacker strategy, we consider an attacker passing through the WSN along a line segment, reaching the base station and leaving the WSN on the opposite direction. We assume an attacker can compromise the sensor nodes located nearby his or her trajectory.

The percentage sensor nodes ordered by distance to the trajectory is parameterized by s_5. Our test case and an example of an attacker passing through the WSN from left side to the right side through the base station is depicted in Fig. 6.

5.3 Number of Deployed Malicious Nodes

The number of malicious sensor nodes in the WSN can vary—based on the options of an attacker—for each deployment strategy. In fact, parameters s_3–s_5 reflect the percentage of deployed malicious sensor nodes for each of the deployment strategy. We analyze the influence of the changed percentage of present malicious sensor nodes from the situation during the optimization in Sect. 7.

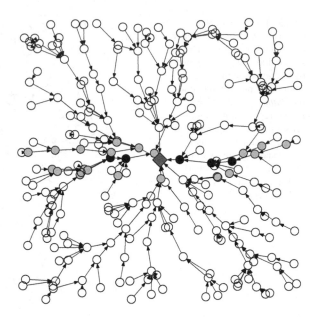

Fig. 6 Topology of the evaluated WSN for the direct center attacker strategy. The sensor nodes are represented by *circles* while the base station is represented by the *red diamond*. The *black circles* represent malicious sensor nodes for the scenario with 2 % malicious sensor nodes and together with the *gray circles* for the scenario with 10 % malicious sensor nodes

5.4 Robustness Evaluation of Optimized Solutions on Different Attacker Strategies

One of the contributions of this chapter is an optimization of the IDS parameters for different attacker strategies. The problem is that the concrete malicious node deployment cannot be reliably predicted in advance. In order to discuss the robustness of the found optimized solutions, we evaluate the performance of optimized IDS for a given attacker strategy against other attacker strategies in Sect. 7. We use two approaches of the evaluation of the impact of changes in the attacker strategy on the IDS performance, the description of which follows.

We use the following notation to explain the evaluations we performed to compare IDSs optimized for various attacker strategies:

Notation 7 *The set* $AS = \{as_1, \ldots, as_{n_{as}}\}$ *is a set of all* n_{as} *attacker strategies evaluated in this chapter.*

Notation 8 *The set* $PF = \{pf_1, \ldots, pf_{n_{as}}\}$ *is a set of all Pareto front approximations. Each Pareto front approximation* $pf_i \in PF$ *where* $1 \le i \le n_{as}$ *was optimized for an attacker strategy* as_i.

Single Pareto Front Approximation Evaluation on Multiple Attacker Strategies
For each Pareto front approximation $pf_i \in PF$ optimized for an attacker strategy as_i, we compute the performance of each IDS setting from pf_i for all other attacker strategies $as_j \in AS \setminus \{as_i\}$. This way, we evaluate the changes of the IDSs performances optimized for a specific attacker strategy in a situation where another attacker strategy operates.

Multiple Pareto Front Approximations Evaluation on Single Attacker Strategy
For each attacker strategy $as_i \in AS$, we compare the performances of the IDS settings
in all Pareto front approximations $pf_j \in PF \setminus \{pf_i\}$ with the performances of the
IDS settings in Pareto front approximation pf_i. This way, we evaluate how the IDSs
performs in a situation where another strategy operates comparing to the IDSs that
were optimized for that situation.

6 Experiment Settings

In this section, we describe experiment settings and optimization scenarios that we
use for evaluation of our IDSs. We also present the ranges of IDS parameters.

6.1 Application

The simulated WSN that we evaluate in this work is inspired by the police unit
scenario in [19]. Each sensor node sends "still alive" packets every second. These
packets can be either dropped or delayed by malicious sensor nodes. The main goal of
our optimization framework is to optimize the IDS for a given scenario (application,
topology, environment, etc.) and to be robust for various attacker strategies in that
environment. We do not aim to provide general IDS setting for any WSN.

One hour of the WSN operation time is simulated in all evaluations.

6.2 Topology and Routing

We build on the topology and routing same as in [12], so that we are able to compare the collaborative and non-collaborative IDS results. The network consists of
250 uniformly distributed sensor nodes deployed in an area of 200 m × 200 m. The
average area for one node is 160 m^2 and the distance between two nearest neighbors
is 12.65 m on average. During the simulation, a node $b_j \in C \cup A$ has 41 neighbors
(nodes from which b_j heard at least one packet during the simulation) on average.

The routing tree is static with longest branches of 8 hops. The topology and the
routing tree are depicted in Figs. 4, 5 and 6.

6.3 IDS Parameters and Sampling

In Table 1, we summarize all eight parameters of the IDS presented in Sects. 3.4–
3.6, their maximum and minimum values. Steps and sampling values are used for
a time demanding sampling that we computed to compare the results found by the
evolution.

Table 1 The list of IDS parameters

Name	Description	Range	Step	Sampling
p1	Maximum monitored nodes	$\langle 1, 30 \rangle$	1/3	3, 9, 27
p2	Buffer size	$\langle 1, 3 \rangle / \langle 1, 10 \rangle$	1	1, 2, 3/3, 6, 9
p3	Minimum received packets	$\langle 1, 30 \rangle$	1/5	1, 15, 30
p4	Detection threshold	$\langle 0.05, 0.95 \rangle / \langle 0.1, 0.9 \rangle$	0.05/0.1	0.25, 0.5, 0.75
p5	Time window	$\langle 10, 300 \rangle$	10/30	10, 150, 300
p6	Minimum received votes	$\langle 1, 10 \rangle$	1/2	1, 5, 10
p7	Voting threshold	$\langle 0.1, 1 \rangle$	0.1	0.25, 0.5, 0.75
p8	Delay timeout	$\langle 1, 5 \rangle$	1	1, 3, 5

If multiple values are presented and divided by "/", the first values were used for the detection of the selective forwarding attack and the second values then for the delay attack detection

Sampling In order to show that evolution can find good enough results in reasonable time, we compared the results found by MOEAs with a true Pareto front found using exhaustive search on multiple computers in [12]. However, we are not able to compute all possible settings for this more complex IDS with additional parameters even if we can run about 200 simulations in parallel in our computational cluster. The exhaustive search would require 148,770,000 simulation runs for the scenario with the selective forwarding attack and 2,479,500,000 for the scenario with the delay attack if all possible settings would be evaluated. One simulation takes approx. 5–8 min.

We decided to sample the search space in the following way. For each parameter p_i, where $i \in \{1, \dots, 7\}$ for the selective forwarding attack and $i \in \{1, \dots, 8\}$ for the delay attack, we choose three carefully considered "sampling" values presented in Table 1. The selection is based on experience and results obtained during early experiments. Then, we iterate over all parameters $p_1, \dots, p_7$ for the selective forwarding attack, respective $p_1, \dots, p_8$ for the delay attack. For each of the parameters, we evaluate all settings within their ranges using steps provided in Table 1. For each value of each parameter p_i, we evaluate all "sampling" settings of all other parameters. Using this approach, we reduce the number of simulations to 84,564 for the selective forwarding attack and to 122,472 for the delay attack.

Having the set of solutions obtained from the aforementioned sampling, we extract only those solutions that are not strictly dominated by any other solutions within this set. We call this extracted set the *Sampling Pareto front approximation*. This set is compared to Pareto front approximations found by evolutions. Note that finding the *Sampling Pareto front approximation* is much more computationally demanding than finding Pareto front approximations using evolution as discussed in Sect. 7. We computed the sampling for an attacker strategy with randomly deployed 5 (2%) malicious sensor nodes for both selective forwarding and delay attacks.

6.4 Evaluated Attacker Strategies

In all experiments, the malicious sensor nodes drop randomly 50% of packets that should be forwarded for the selective forwarding attack. For evaluation of the delay attack, the malicious sensor nodes delay all packets within interval of $\langle 0, 5 \rangle$ s.

Deployment of Malicious Nodes We optimize the IDS for six different deployments of malicious sensor nodes for the selective forwarding attack. For each deployment strategy (random, center drop and direct centre), we evaluate two cases: with 5 malicious sensor nodes (2%) and with 25 malicious sensor nodes (10%). All attacker strategies are illustrated in Figs. 4, 5 and 6.

For the delay attack, we evaluate only the case with 5 (2%) randomly deployed malicious sensor nodes due to computational time restrictions. However, the characteristics that are related to overhearing the communication and to the collaborative decision are equivalent to the selective forwarding attack that we evaluated thoroughly.

No "leaf" sensor node (a node that has no descendant) is malicious within the experiments. These sensor nodes can neither perform efficiently the selective forwarding or the delay attack (no packets are forwarded by them), nor can be detected by any IDS node (no packets addressed to them can be overheard).

7 Experiment Results

In this section, all experiment results are presented and discussed. First, we show the increased accuracy of the collaborative IDS in comparison with the non-collaborative IDS, in the case of detecting the selective forwarding attack. Then we compare evolution performance for selective forwarding and delay attacks with much more time demanding sampling. Finally, we provide a mutual comparison of IDS settings optimized for various attacker strategies.

7.1 Selective Forwarding Attack

First, we briefly compare the Pareto front approximation found by sampling for the collaborative IDS with a true Pareto front found for the non-collaborative IDS in [12]. Both detection techniques are evaluated in the same WSN and attacker strategy—random deployment of 2% malicious sensor nodes. In Fig. 7, we show different views

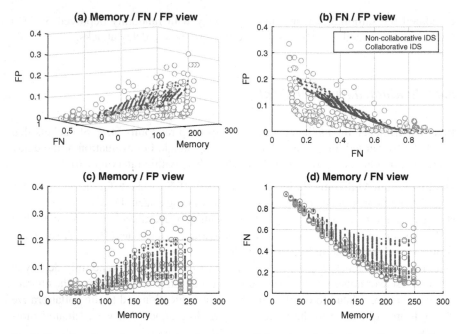

Fig. 7 Sampling Pareto front approximation for the collaborative selective forwarding attack detection compared to the true Pareto front for the non-collaborative selective forwarding attack detection. All the solutions found by sampling for the collaborative detection dominate the solutions found by exhaustive search for the non-collaborative detection

of the optimized solutions for both detection techniques in the three-dimensional objective space.

In Fig. 7a, we show that all sampled non-dominated solutions found for collaborative IDS dominate the Pareto optimal solutions found for non-collaborative IDS. Measuring the dominated volume of the objective space by the hyper-volume indicator, we obtain 0.525 for the non-collaborative IDS and 0.574^2 for the collaborative IDS. In Fig. 7b, we show that the collaboration among the IDS nodes can significantly decrease the number of false positives—a consensus has to be made to label a node as attacker. A decrease of false negatives is caused by dividing the monitoring time into smaller windows, where, in each of them, a potential dropping can be detected. In Fig. 7c, d, we can see that higher memory consumption caused particularly by a higher number of monitored neighbors decreases false negatives on one hand (more neighbors being monitored means also higher number of truly recognized malicious nodes), but increases false positives on the other hand (if a neighbor is not monitored, it can neither be labeled as malicious one truly, nor falsely).

As we shown, the collaborative approach provides better IDS results even though we are not able to evaluate the whole search space. However, the collaborative IDS

---

[2] As shown below, evolution can improve the results farther.

requires a communication overhead. Nevertheless, we would like to emphasize that the overhead is not significant—at least for our simulated application. Each sensor node $c_k \in C$ initiates the voting scheme at most once for each monitored neighbor $b_l \in A \cup C$ during the whole WSN operation time. Each neighbor $c_i \in C$ of the c_k has to answer to the voting request if the b_l is also monitored by the c_i. Sensor nodes monitor 30 neighbors at most and are monitored by at most 30 neighbors on average. That means that c_k initiates 30 voting request at most in case all monitored neighbors are suspicious. The average sensor node c_k has to answer on 30 voting requests (but only in the unrealistic worst case—only if it also monitors the sensor node in the request) on average for each of 41 neighbors (see Sect. 6.2 for more details). The overall overhead would be 1260 packets sent by each IDS node in such very unrealistic worst case. Note that the sensor nodes in a distance of one hop from the base station forward approx. 2500 packets during only one hour of the operation time of the WSN.

Evolution can speedup the process of finding solutions that are similar or even dominate the solutions found by the sampling. We present results of two multiobjective evolution runs (marked as "Evo #1" and "Evo #2"). The evolution settings are described in Sect. 4.2. The solutions found by the evolution compared to those found by sampling are depicted in Fig. 8.

In Table 2, we present results of all four metrics for each of the optimization process. We can see that both evolution runs outperforms the sampling according to

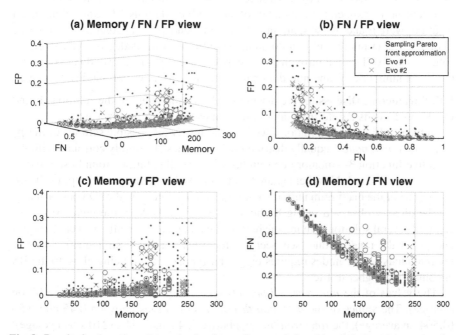

Fig. 8 Results for detection of the selective forwarding attack found by evolution compared to the results of the Sampling Pareto front approximation

Table 2 Performance metrics of the optimization of the selective forwarding attack

	Hyper-volume	Coverage			Simulations	Solutions
		Evo #1	Evo #2	Sampling		
Evo #1	0.577	–	65 % (94/144)	100 % (144/144)	13,502	144
Evo #2	0.583	82 % (126/153)	–	98 % (150/153)	23,558	153
Sampling	0.574	22 % (45/201)	19 % (39/201)	–	84,564	201

Hyper-volume indicator—result of the hyper-volume indicator for each optimization process. *Coverage metric*—for each Pareto front approximation in a row, the values specify the number of found solutions that are not dominated by any solution within the result set of Pareto front approximation specified in the column. *Number of simulations*—number of simulation runs within the whole optimization process. *Number of non-dominated solutions*—number of resulting non-dominated solutions

the hyper-volume indicator. Measuring the performance by mutual coverage of the solutions, we can see that solutions found by the evolution runs dominate nearly all solutions found by the sampling. However, we were able to find several solutions that have a lower number of false negatives using the sampling than any solution found by the evolution (see Fig. 8b). The process of optimization by the evolution was less time demanding as declare the numbers of required simulations. More non-dominated solutions were found by the sampling. We recommend to use a larger population size to obtain higher number of non-dominated solutions, if needed. However, since the solutions are well spread through the objective space (see Fig. 8), we do not consider the lower number of found non-dominated solutions as an important disadvantage.

IDS Parameters Discussion We discuss the IDS parameters of the solutions found by "Evo #2" as its Pareto front approximation evinces best performance.

In the resulting set of solutions, we can find nearly all possible settings of the p_1 (*max monitored nodes*) equally distributed. The values absolutely correlate with the objective function 3—memory consumption, since all solutions found use a *buffer size* (p_2) equaled to 1. No other parameter influences the memory consumption. Thus, the impact of the maximum number of monitored neighbors on the IDS performance can be directly observed in Fig. 8 (the axis "Memory"). W.r.t. the *buffer size*—we can find several solutions having the buffer size equaled to 2 in all attacker strategies except for the random one (see Sect. 7.3 for other evaluated strategies). Some of the malicious sensor nodes requires bigger buffer size due to being close to the BS encountering higher traffic.

The values of the *min received packets* (p_3) varies between 1 and 11 (5.07 on average). The values of the *detection threshold* (p_4) varies between 0.45 and 0.55 (0.504 on average), the *time window* (p_5) between 43 and 294 s (210 s on average), the *minimum received votes* (p_6) between 1 and 7 (2.79 on average) and the voting threshold between 0.28 and 0.99 (0.86 on average).

7.2 Delay Attack

In this section, we present non-dominated results for the detection of the delay attack found both by the sampling and evolution. In Fig. 9, we compare the performance of the IDS settings found by both the evolution and the sampling in the objective space. The number of non-dominated solutions found by sampling is reduced comparing to the selective forwarding attack. It may be caused by the fact that we are not able to compute such "dense" sampling—the search space is much larger (see Table 1). Since the basic principle of the delay attack detection is similar to the selective forwarding, we can observe similar patterns of the Pareto front approximations. Main difference is higher memory consumption needed to obtain comparable false negatives. This is caused by a need of storing the packets in the IDS buffer for longer time.

In Table 3, we can see that the evolution provides better results than the sampling w.r.t all the metrics—similarly to the selective forwarding attack.

IDS Parameters Discussion Such as for the selective forwarding attack, results found by "Evo #2" are discussed for the delay attack.

The number of *max monitored nodes* (p_1) varies between 1 and 24 (15.9 on average). The *buffer size* (p_2) varies between 1 and 9 (2.3 on average) and only the solutions with p_1 higher than 18 requires more than 3 slots in the buffer. The parameter *min received packets* p_3 evinces values between 1 and 8 (2.2 on average). The values of the *detection threshold* (p_4) varies between 0.07 and 0.64 (0.38 on average).

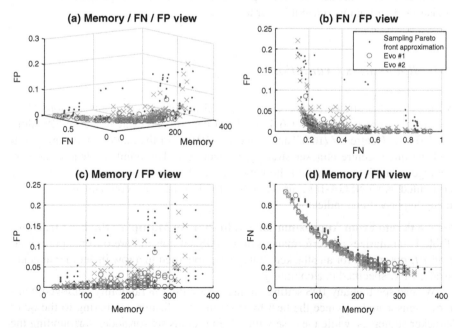

Fig. 9 Results for detection of the delay attack found by evolution compared to the results of the Sampling Pareto front approximation

Table 3 Performance metrics of the optimization of the delay attack

	Hyper-volume	Coverage			Simulations	Solutions
		Evo #1	Evo #2	Sampling		
Evo #1	0.604	–	61 % (72/118)	97 % (115/118)	13,539	118
Evo #2	0.625	79 % (112/141)	–	99 % (139/141)	23,953	141
Sampling	0.582	22 % (30/139)	14 % (19/139)	–	122,472	139

Hyper-volume indicator—result of the hyper-volume indicator for each optimization process. *Coverage metric*—for each Pareto front approximation in a row, the values specify the number of found solutions that are not dominated by any solution within the result set of Pareto front approximation specified in the column. *Number of simulations*—number of simulation runs within the whole optimization process. *Number of non-dominated solutions*—number of resulting non-dominated solutions

Note that the behavior of the malicious nodes is different to the selective forwarding attack—the malicious nodes delay randomly all packets within the interval $\langle 0, 5 \rangle$. The values of *time window* (p_5) varies between 55 and 295 s (131 s on average), the *minimum received votes* (p_6) varies between 1 and 2 (1.23 on average) and the voting threshold varies between 0.17 and 0.99 (0.77 on average). The *delay timeout* (p_8) was set to 1 s in each solution—packets forwarded by benign node are usually transmitted within this timeout in our test case.

7.3 Robustness Evaluation

Various attacker strategies discussed in Sect. 5 are evaluated in this section. Their settings was presented in Sect. 6.4. We label each of the evaluated case "Random 2 %/10 %", "Centre 2 %/10 %" and "Line 2 %/10 %" for the random, the center drop and the direct centre attacker strategy, respectively. The numbers denote the percentage of malicious nodes in the network. All Pareto front approximations were computed using NSGA-II set up according to "Evo #2". We compare all the results using hyper-volume indicator.

Single Pareto Front Approximation Evaluation on Multiple Attacker Strategies
Table 4 summarizes all performances of the IDS settings in Pareto front approximation pf_i optimized for an attacker strategy as_i specified in a row in another attacker strategy as_j specified in a column j.

We can see that any IDS settings in the case with 2 % randomly deployed malicious sensor nodes evince the best hyper-volume indicator comparing to the other attacker strategies, while the case with 2 % malicious sensor nodes surrounding the BS evinces the worst hyper-volume indicator. We found out that while the memory

Table 4 Hyper-volume indicator results for various deployment strategies

	R 2%	R 10%	C 2%	C 10%	L 2%	L 10%	Average diff
Random 2%	**0.583**	0.520	0.453	0.487	0.534	0.462	0.0277
Random 10%	0.567	**0.555**	0.465	0.510	0.540	0.504	0.0107
Centre 2%	0.562	0.539	**0.485**	0.507	0.539	0.495	0.0130
Centre 10%	0.565	0.539	0.473	**0.515**	0.533	0.499	0.0135
Line 2%	0.565	0.494	0.461	0.473	**0.554**	0.406	0.0420
Line 10%	0.567	0.545	0.466	0.510	0.537	**0.513**	0.0112
Average	0.568	0.532	0.467	0.500	0.540	0.480	

For each strategy A in a row, the values specify the hyper-volume indicator of the IDS settings optimized for A in another deployment strategy B specified in a column. Looking to the table from the other perspective, for each strategy A in a column, the values specify the hyper-volume indicator of the IDS settings optimized for a strategy B specified in a row but evaluated in the strategy A. The values in bold states for the best results for each of the strategy—IDS settings optimized and evaluated in the same deployment strategy. The last column specifies the average difference of an IDS optimized for an attacker strategy in a row to the best result achieved for each of the strategy in the column

consumption is constant and the number of false positives do not change significantly, more significant changes in the number of false negatives can be observed. This is caused by the placement of malicious nodes—it is more difficult to detect a malicious sensor node surrounded by other malicious sensor nodes or a malicious node close to the edge of the WSN receiving packets from only one descendant.[3]

See Fig. 10 for an example of IDS optimized for Random 2% in the attacker strategy Random 10% (a–c) and vice versa (d–f). While the memory consumption is constant and the number of false positives do not change significantly, we can see more significant changes in the number of false negatives due to higher number of "close-to-edge" sensor nodes.

Multiple Pareto Front Approximations Evaluation On Single Attacker Strategy
In Table 4, we can see for any attacker strategy as_j in a column j how each hyper-volume indicator of the IDS settings pf_i optimized for each strategy as_i in a row i differs to the hyper-volume indicator of the IDS settings pf_j evaluated (and also optimized) in the as_j. In the last column of the Table 4, we present the average difference to the best value for each Pareto front approximation pf_i across all attacker strategies. The lower the average difference, the more robust is the Pareto front approximation in another attacker strategies.

We can see that Pareto front approximation of IDS settings optimized for attacker strategy "Random 10%" performs the best across all other strategies followed by

---

[3]Such traffic can be overheard by less (if any) number of neighbors comparing to a sensor node placed closer to the BS receiving packets from several directions.

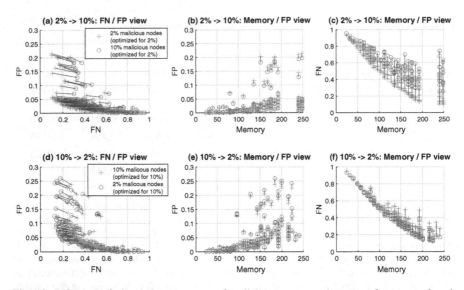

Fig. 10 Influence of changed percentages of malicious sensor nodes on performance of each optimized IDS setting in the random attacker strategy. Green crosses represent IDS performance in an environment for which the IDS was optimized (2 % for Figures (**a–c**) and 10 % for Figures (**d–f**)). *Red circles* represent IDS performance in an environment with increased [Figures (**a–c**) and decreased (Figures (**d–f**)] percentage of malicious sensor nodes. *Lines* connect equal IDS settings

"Line 10 %". On the other hand, the solutions found for the attacker strategy "Line 2 %" performs the worst on average in the other strategies.

8   Conclusion

We proposed and implemented a complex optimization framework consisting of an optimization engine and a simulator. The simulations can be executed on multiple computers in a distributed manner. This optimization framework is aimed at but not limited to optimization of intrusion detection systems to detect different types of active attacks on a WSN. In the simulator, the target WSN can be specified in details including the environment, topology, physical properties of the sensor nodes, routing and application.

In this work, we demonstrated usability of the optimization framework on the selective forwarding attack and the delay attack detection. We have shown that efficient Pareto front approximation can be found using multi-objective evolutionary algorithm in a reasonable time. Four different metrics were used to evaluate the optimization processes. The diversity of the non-dominated solutions can provide a network operator with an option to choose any solution according to requirements that can be changed during the WSN operation time.

We discussed thoroughly attacker strategies of the selective forwarding and the delay attack as well as usability of our detection techniques for each variation of any attacker strategy. The IDS was optimized to six different deployments of malicious sensor nodes and the resulting non-dominated IDS solutions were evaluated for robustness on each of the deployment.

The optimization framework can be used directly for, e.g., IDS that we implemented within our security middleware for WSNs—"WSNProtectLayer" [19].

Acknowledgements We would like to thank Ludek Smolik, Lukas Sekanina and colleagues from CRoCS for the discussions and suggestions. This work was supported by the Czech research Project VG20102014031, programme BV II/2—VS. Access to computing and storage facilities owned by parties and projects contributing to the National Grid Infrastructure MetaCentrum, provided under the programme "Projects of Large Research, Development, and Innovations Infrastructures" (CESNET LM2015042), is greatly appreciated.

References

1. da Silva, A.P.R., Martins, M.H.T., Rocha, B.P.S., Loureiro, A.A.F., Ruiz, L.B., Wong, H.C.: Decentralized intrusion detection in wireless sensor networks. In: Proceedings of the 1st ACM International Workshop on Quality of Service & Security in Wireless and Mobile Networks, pp. 16–23 (2005)
2. Stehlik, M., Matyas, V., Stetsko, A.: Towards better selective forwarding and delay attacks in wireless sensor networks. In: Proceedings of the 13th IEEE International Conference on Networking, Sensing, and Control (2016)
3. Karlof, C., Wagner, D.: Secure routing in wireless sensor networks: attacks and countermeasures. AdHoc Netw. J. 1(2), 293–315 (2003)
4. Krontiris, I., Dimitriou, T., Freiling, F.C.: Towards intrusion detection in wireless sensor networks. In Proceedings of the 13th European Wireless Conference (2007)
5. Tiwari, M., Arya, K.V., Choudhari, R., Choudhary, K.S.: Designing intrusion detection to detect black hole and selective forwarding attack in WSN based on local information. Proceedings of the 2009 Fourth International Conference on Computer Sciences and Convergence Information Technology. ICCIT '09, pp. 824–828. IEEE Computer Society, Washington, DC (2009)
6. Hai, T.H., Huh, E.: Detecting selective forwarding attacks in wireless sensor networks using two-hops neighbor knowledge. In: Seventh IEEE International Symposium on Network Computing and Applications, pp. 325–331 (2008)
7. Liu, F., Cheng, X., Chen, D.: Insider attacker detection in wireless sensor networks. In: INFO-COM 2007. 26th IEEE International Conference on Computer Communications, pp. 1937–1945. IEEE (2007)
8. Khanna, R., Liu, H., Chen, H.H.: Self-organization of sensor networks using genetic algorithms. In: IEEE International Conference on Communications, 2006. ICC'06, vol. 8, pp. 3377–3382 (2006)
9. Khanna, R., Liu, H., Chen, H.H.: Dynamic optimization of secure mobile sensor networks: a genetic algorithm. In: IEEE International Conference on Communications, 2007. ICC'07, pp. 3413–3418, (2007)
10. Khanna, R., Liu, H., Chen, H.H.: Reduced complexity intrusion detection in sensor networks using genetic algorithm. In: IEEE International Conference on Communications, 2009. ICC'09, pp. 1–5 (2009)
11. Heady, R., Lugar, G., Servilla, M., Maccabe, A.: The Architecture of a Network Level Intrusion Detection System. Technical report, University of New Mexico, Albuquerque, NM (1990)

12. Stehlik, M., Saleh, A., Stetsko, A., Matyas, V.: Multi-objective optimization of intrusion detection systems for wireless sensor networks. In: Li, P., et al. (eds.) Advances in Artificial Life, ECAL 2013, Proceedings of the Twelfth European Conference on the Synthesis and Simulation of Living Systems, pp. 569–576. MIT Press, Cambridge, MA (2013)
13. Banerjee, S., Grosan, C., Abraham, A.: IDEAS: intrusion detection based on emotional ants for sensors. In: Proceedings of 5th International Conference on Intelligent Systems Design and Applications, 2005. ISDA '05, pp. 344–349. IEEE (2005)
14. Banerjee, S., Grosan, C., Abraham, A., Mahanti, P.K.: Intrusion detection on sensor networks using emotional ants. Int. J. Appl. Sci. Comput. **12**(3), 152–173 (2005)
15. Mukherjee, P., Sen, S.: Using learned data patterns to detect malicious nodes in sensor networks. In: Proceedings of the 9th International Conference on Distributed Computing and Networking. ICDCN'08, pp. 339–344. Springer, Berlin (2008)
16. Roosta, T., Shieh, S., Sastry, S.: Taxonomy of security attacks in sensor networks and countermeasures. In: The First IEEE International Conference on System Integration and Reliability Improvements, vol. 25, p. 94 (2006)
17. Loo, C.E., Ng, M.Y., Leckie, C., Palaniswami, M.: Intrusion detection for routing attacks in sensor networks. Int. J. Distrib. Sens. Netw. **2**(4), 313–332 (2006)
18. Stetsko, A., Smolka, T., Matyas, V., Stehlik, M.: Improving intrusion detection systems for wireless sensor networks. In: Boureanu, I., et al. (eds.) Applied Cryptography and Network Security. Lecture Notes in Computer Science, vol. 8479, pp. 343–360. Springer, Berlin (2014)
19. Matyas, V., Svenda, P., Stetsko, A., Klinec, D., Jurnecka, F., Stehlik, M.: Securing Cyber Physical Systems, chapter 5: WSNProtectLayer Security Middleware for Wireless Sensor Networks. CRC Press, Boca Raton, FL (2015). ISBN 978-1-4987-0098-6
20. Roman, R., Lopez, J., Gritzalis, S.: Situation awareness mechanisms for wireless sensor networks. IEEE Commun. Mag. **46**(4), 102–107 (2008)
21. Anderson, D.P.: BOINC: a system for public-resource computing and storage. In: Proceedings of IEEE/ACM Workshop on Grid Computing, pp. 4–10 (2001)
22. Köpke, A., Swigulski, M., Wessel, K., Willkomm, D., Klein Haneveld, P.T., Parker, T.E.V., Visser, O.W., Lichte, H.S., Valentin, S.: Simulating Wireless and Mobile Networks in OMNeT++ the MiXiM Vision. In: Proceedings of the 1st International Conference on Simulation Tools and Techniques for Communications, Networks and Systems & Workshops, Simutools '08, pp. 71–78, ICST (Institute for Computer Sciences, Social-Informatics and Telecommunications Engineering), Brussels (2008)
23. OMNeT++. OMNeT++ Network Simulation Framework—Homepage. http://www.omnetpp.org/. Accessed 22 Oct 2015
24. Stetsko, A., Stehlik, M., Matyas, V.: Calibrating and comparing simulators for wireless sensor networks. In Proceedings of the 8th IEEE International Conference on Mobile Adhoc and Sensor Systems, pp. 733–738. Los Alamitos (2011)
25. Rappaport, T.: Wireless Communications: Principles and Practice, 2nd edn. Prentice Hall PTR, Englewood Cliffs, NJ (2001)
26. Crossbow. TelosB Datasheet. http://www.willow.co.uk/TelosB_Datasheet.pdf. Accessed 26 Oct 2015
27. Talbi, E.G.: Metaheuristics—From Design to Implementation. Wiley, New York (2009)
28. Deb, K., Pratap, A., Agarwal, S., Meyarivan, T.: A fast and elitist multiobjective genetic algorithm: NSGA-II. IEEE Trans. Evol. Comput. **6**(2), 182–197 (2002)
29. Zitzler, E., Laumanns, M., Thiele, L.: SPEA2: Improving the Strength Pareto Evolutionary Algorithm. Technical report, Eidgenössische Technische Hochschule Zürich (ETH) (2001)
30. Auger, A., Bader, J., Brockhoff, D., Zitzler, E.: Theory of the hypervolume indicator: optimal μ-distributions and the choice of the reference point. In: Proceedings of the Tenth ACM SIGEVO Workshop on Foundations of Genetic Algorithms. FOGA '09, pp. 87–102. ACM. New York, NY (2009)
31. Fonseca, C.M., Paquete, L., Lopez-Ibanez, M.:. An improved dimension-sweep algorithm for the hypervolume indicator. In: IEEE Congress on Evolutionary Computation, 2006. CEC 2006, pp. 1157–1163 (2006)

32. Zitzler, E., Thiele, L.: Multiobjective evolutionary algorithms: a comparative case study and the strength Pareto approach. IEEE Trans. Evol. Comput. **3**(4), 257–271 (1999)
33. Jurnecka, F., Stehlik, M., Matyas, V.:. On node capturing attacker strategies. In: Security Protocols XXII—22nd International Workshop Cambridge. Revised Selected Papers, pp. 300–315. Springer LNCS (2014)
34. Yu, B., Xiao, B.: Detecting selective forwarding attacks in wireless sensor networks. In 20th International Parallel and Distributed Processing Symposium, 2006. IPDPS 2006. IEEE (2006)

Computational Intelligence Based Security in Wireless Sensor Networks: Technologies and Design Challenges

Pratik Ranjan and Hari Om

Abstract The wireless sensor networks (WSNs) are emerging as a fast moving technology for future computing solutions. However, there is a major issue related to security in WSN environments. The application of intelligent computing works efficiently for many computing and scientific problems; but, the computational intelligence (CI) based security schemes have not been properly explored. Here, we explore the CI-based approach for scientific computing problems and try to find the recent challenges and future opportunities for developing the CI-based security schemes for WSNs.

Keywords Computational intelligence · Fuzzy logic · Neural networks · Security · Wireless sensor networks

1 Introduction

The wireless sensor networks (WSNs) are event-monitoring and data collecting devices, which work as an interface between a physical environment and a computer system. They play a major role in gathering the data from environment where the pure wired connection is not so easy to establish and communicate that data to a centre where it can proceed further. The sensor networks are generally deployed for periodic reporting and event detection in an environment [35]. However, the sensor nodes in a WSN have limited battery power, less storage capacity and less computation capacity, which may lead to node failure. Apart from the limited capabilities of sensor nodes, there are numerous challenges in a WSN like design and deployment of sensor nodes, mobility and topology changes, localization and physical distribution, clustering, data aggregation, security, and quality of service management. The secu-

P. Ranjan (✉) · H. Om
Department of Computer Science and Engineering, Indian Institute of Technology (Indian School of Mines), Dhanbad, India
e-mail: pratik.ismdhanbad@gmail.com

H. Om
e-mail: hariom4india@gmail.com

© Springer International Publishing AG 2017   131
A. Abraham et al. (eds.), *Computational Intelligence in Wireless Sensor Networks*,
Studies in Computational Intelligence SCI 676, DOI 10.1007/978-3-319-47715-2_6

rity is a very important factor in any communication or network device. In case of the WSNs, it is more demanding and challenging because of the constrained battery and memory capacity. There are many security issues that need to be addressed in WSNs like discovery and verification of nodes, key establishment, secure routing, node authentication, secure group management and secure data aggregation [29, 64].

The policy-based approach in a network handles the challenges quite well in a static environment, but the sensor nodes are deployed in dynamic environments where an intelligent-based approach works more efficiently than the policy-based approach. The intelligence enables each host to learn new states, events, and actions so that the optimal or near-optimal action can be decided. The most common computational intelligence (CI) paradigms are fuzzy systems, artificial neural networks, evolutionary computation, swarm intelligence, and artificial immune systems. Researchers have proposed many CI-based schemes to resolve the challenging issues in WSNs like node coverage and energy management [20, 25], localization and optimization [22], but the CI-based security schemes are not so popular especially for WSNs. The fuzzy based security schemes in WSNs have been presented in [30, 81]; however, the other approaches like genetic algorithm [6, 83], swarm intelligence, and artificial neural network based security schemes for WSNs are in their infancy stage of development.

This leads to motivate us to study the current trends in CI-based schemes in different areas including WSNs and try to figure out the challenges and future opportunities for the development of CI-based secure schemes for WSNs. This work will help to find the CI-based security challenges and their probable solutions for dynamic as well as infrastructure-less network environments.

2   Wireless Sensor Networks

A wireless sensor network (WSN) is a collection of sensor nodes that collectively work for some particular task like weather forecasting, event detection, intrusion detection, health and area monitoring, etc. In a WSN, each sensor node consists of one or more sensing devices that communicate to few other local sensor nodes via wireless channels. There are three major limitations in a sensor node, namely, battery power, communication bandwidth, and storage capacity. The WSNs support a number of real-world applications that lead to a challenging research and engineering problem because of the flexibility and dynamic property of sensor nodes. Accordingly, there is no single set of requirements that clearly classifies all WSNs, and also there is not a single technical solution that encompasses the entire design space. Many of these applications share some basic characteristics. In most of the WSNs, the sources of data are the actual nodes that sense the data and the sink nodes are the delivery nodes of final data.

2.1 Types of WSNs

Different types of Wireless Sensor Networks (WSNs) as listed in [1] are:

(i) **Terrestrial WSNs** The terrestrial WSNs consist of huge collection of sensor nodes either in an unstructured or a structured manner for efficient communication with the base station. In unstructured mode, the sensor nodes are randomly distributed within the target/monitoring area. The pre-planned or structured mode considers optimal placement of the sensor nodes.

(ii) **Underground WSNs** The underground WSNs are more expensive than the terrestrial WSNs in terms of deployment, maintenance, and planning. These WSNs are hidden in ground to monitor the underground conditions. Additional sink nodes are located on or above the ground to relay the information from the sensor nodes to the base station. The underground environment makes wireless communication a challenge due to high level of attenuation and signal loss.

(iii) **Underwater WSNs** The underwater WSNs consist of a number of sensor nodes and vehicles deployed under the water. The autonomous underwater vehicles are used for gathering the data from these sensor nodes. A challenge of underwater communication is long propagation delay, bandwidth, and node failures. The issue of energy conservation for underwater WSNs involves the development of underwater communication and networking techniques.

(iv) **Multimedia WSNs** The objective of these WSNs is to enable tracking and monitoring of events in the form of multimedia data. These networks consist of low-cost sensor nodes equipped with microphones and cameras.These nodes are interconnected with each other over a wireless connection for data compression, data retrieval, and correlation. The challenges in the multimedia WSNs include high energy consumption, high bandwidth requirements, data compression, and data processing.

(v) **Mobile WSNs** The mobile WSNs consist of a collection of sensor nodes that can move their own and they can be interacted with the physical environment. The mobile nodes have an ability to sense, compute, and communicate the data. The mobile WSNs are much more versatile than the static sensor networks. The advantages of these sensors over the static one include better and improved coverage, better energy efficiency, and superior channel capacity.

2.2 Applications of WSNs

There are several applications of the WSNs like industrial control systems, event detection, health monitoring, environmental monitoring, battlefield surveillance, object monitoring including tracking the movements and patterns of objects, insects, or animals. The WSNs can be deployed in mission critical applications such as security of key land marks, surveillance of buildings and bridges, etc. Depending on the application challenges and constraints, the WSNs can adopt different forms, use different technologies, and communicate through different network topologies [3].

2.3 Security Issues in WSNs

The WSNs have many security challenges due to their inherent limitations in communication bandwidth. The hardware and software developments have addressed these issues to some extent, but a complete secure sensor network requires deployment of countermeasures such as node authentication, secure routing, secure key management, and lightweight encryption techniques [11, 51].

The common attacks in WSNs [8, 18, 51, 58] are as follows:

- **Spoofed Attack** It is a kind of replay routing or altering attack where an adversary may send a false error message, create routing loops or change the routes for packet delivery.
- **Selective Forwarding Attack** This is a black hole type of attack where an adversary gets the control of communication channel and may drop the packet or refuse to forward any particular packet to the next node.
- **Sinkhole Attack** In this attack, an imaginary high quality route is advertised by the adversary on which all the nodes start routing. Because of this sinkhole formation, all nodes start routing through a compromised node.
- **Sybil Attack** In this attack, an adversary presents his identity in multiple forms so that it appears and functions as multiple distinct nodes. After becoming a part of the network, the adversary node may overhear the communications or may control the network traffic.
- **Wormhole Attack** In this type of attack, an adversary collect the packets at one point, tunnel them to some other point and further replay those packets into networks. Through wormhole attack, without any knowledge of the secret key, the attacker can relay the authentication exchanges to gain access.
- **Hello Flood Attack** A sensor node uses *Hello* message for detecting the neighbouring nodes. The same strategy is used by an adversary by sending the fake *Hello* message to a number of nodes. The motive of the attacker in this type of attack is to compromise every neighbouring node.
- **Denial of Service (DoS) Attack** In this type of attack, an adversary sends a number of false packets to a particular network with an objective to create a flood of messages in the network. For WSNs that have limited battery and storage capacity, the DoS attack creates a shortage of both the constraints and makes the system inactive or out-of-order.
- **Acknowledge Spoofing Attack** In this type of attack, an adversary acts as a man-in-the-middle and sends the false acknowledgment message to the sender on behalf of a genuine receiver node.

The challenges for information processing in WSNs are design methodologies and tools to support distributed signal processing, secure data processing, networking, information storage and management, and application development. The key management and secure routing protocols are the resilient solutions for the attacks identified so far. However, the research towards a complete secure sensor network is still in its development stage. For designing a new security scheme, the battery energy, memory space, and the cost of sensor nodes must be considered appropriately by the researchers. These parameters (battery energy, memory space, and cost of sensor node) lead the researchers to develop a new scheme that will be secure and efficient enough to implement in sensor nodes.

3 Computational Intelligence

Computational Intelligence (CI) is a study of adaptive mechanisms to enable or facilitate intelligent behavior in complex and changing environments [19]. It is a sub-branch of Artificial Intelligence (AI) that mainly focuses on those AI paradigms that exhibit an ability to learn and adapt to new situations, to generalize, abstract, and discover [5].

3.1 Types of Computational Intelligence Techniques

There are several computational intelligence (CI) paradigms based on which different CI techniques may be classified. The neural networks, genetic algorithms, fuzzy systems, evolutionary computing and artificial life are the five basic building blocks of CI [49]. The early definition of CI included fuzzy sets, neural networks, genetic algorithms, and probabilistic reasoning. The modern definition of CI [32] includes biologically inspired model of machine intelligence like granular computing, evolutionary computing and their interactions with artificial life, chaos theory, etc. Figure 1 presents a list of modern CI techniques.

3.1.1 Granular Computing

The granular computing (GC) is a new paradigm of information processing that deals with collection of entities arranged together according to their similarity, functional adjacency, indistinguishability, coherency or alike [4]. Formation of information granules can be done by two processes: (i) data abstraction, and (ii) derivation of knowledge from information. Information granulation supports communications at different levels: between humans, humans and computers, computers and computers. Numerical computing is data-oriented whereas the GC is knowledge-oriented. The most popular models for granular computing are as follows:

Fig. 1 Modern
Computational Intelligence
Paradigms

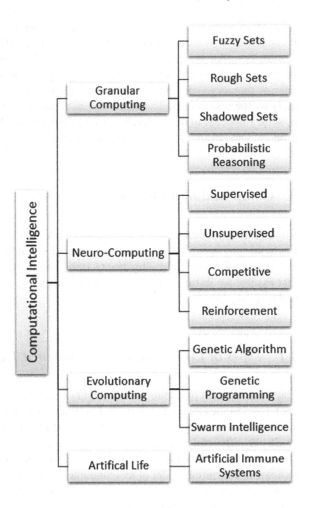

- **Fuzzy Sets** Classical set theory allows an element to be either included in a set or excluded, whereas the fuzzy sets allow an object to be a partial member of a set. The fuzzy rules are of general form as if antecedent(s) then consequent(s), where antecedents and consequents are propositions containing linguistic variables. The antecedent of a fuzzy rule forms a combination of fuzzy sets using logic operations. So, the fuzzy sets and fuzzy rules together form the knowledge base of a rule-based inference system [35].
- **Rough Sets** Rough set theory [60] has been conceived as a tool to conceptualize, organize, and analyze various types of data to deal with inexact, uncertain or vague knowledge in applications related to artificial intelligence [33]. It is a framework in which we represent the concepts in the setting of indiscernibility relations. The application of rough set theory is to learn classification rules.

Though the fuzzy sets and rough sets both use membership functions, yet they are different in implementation. In rough sets, the lower and upper approximation of the rough set is determined. The lower approximation consists of all elements that belong with full certainty to the corresponding set whereas the upper approximation consists of elements that may possibly belong to the set [19]. The rough sets are frequently used in machine learning as classifier and in extracting knowledge from the incomplete data.

- **Shadowed Sets** Shadowed sets are directly induced by the fuzzy membership functions and they are designed to conserve the amount of uncertainty in the original fuzzy sets. They reveal interesting conceptual and algorithmic relationships existing between rough sets and fuzzy sets. In shadowed sets, three quantification levels describing the elements of the set 0, 1, and [0, 1] are used to simplify the fuzzy relation. Conceptually, the shadowed sets are close to rough sets even though their mathematical foundations are very different [62]. The concepts of negative region, lower bound, and boundary region in rough set theory correspond to three-logical values 0, 1 and [0, 1]. These values are named as excluded, included, and uncertain, respectively. The shadowed sets can be considered as a bridge between the fuzzy and rough sets [85].

- **Probabilistic Reasoning** "Probability is not really about numbers; it is about the structure of reasoning" [61]. The motive of probability theory is to provide a logical view on uncertain and partial information. The probabilistic relationships can be represented by *Bayesian Networks* that provide a clear visual representation for many independent relationships embedded in a probabilistic model.

3.1.2 Neuro Computing

It is a branch of computing in which we study the biological neural networks and try to construct a similar framework with some mathematical model for machine learning applications. The *Artificial Neural Networks* (ANN) are electrical analogue of the biological nervous system [32]. The five popular ANN based machine learning techniques are given below:

- **Supervised Learning** It is a training-based learning method in which the algorithm adapts the weights or thresholds of the neurons to produce the desired output instances on a given input instance of the network. Here the algorithm is initialized with a random set of weights and thresholds. After each iteration, the algorithm evaluates the errors at the output of each of the neurons and tries to adapt the weights that minimize the error. Some of the popular models of supervised learning are McCulloch and Pitts Model [52], Perceptron Model [53, 72] and Backpropagation Method [12].

- **Unsupervised Learning** It does not require any trainer and has no fixed target output. It uses a feature selection based classification technique for selecting the objects. Unlike supervised learning which is based on the feed forward geometry, the unsupervised learning allows propagation of signal both in forward and backward directions.

- **Competitive Learning** Competitive learning is a kind of learning in which the neurons compete to each other on a combination network of feed-forward and feedback for the desire output.
- **Reinforcement Learning** It is a neurodynamic-programming model which is neither supervised nor purely unsupervised. It is a CI based approach in which a system learns to choose optimal actions in a dynamic environment [78]. Here an agent acts on the environment to cause a transition of environmental state and receives an immediate reward for its action [26, 77].

3.1.3 Evolutionary Computing

Evolutionary computing (EC) is based on the concept of "survival of the fittest: the weak must die". Engelbrecht [19] describes the evolutionary algorithms as follows:

It uses a population of individuals known as a chromosome that defines the characteristics of individuals in the population. Each characteristic is referred to as a gene and the value of a gene is referred to as an allele. For each generation, individuals compete to reproduce offspring. Those individuals with the best survival capabilities have the best chance to reproduce. Offspring are generated by combining parts of the parents, a process referred to as crossover. Each individual in the population can also undergo mutation which alters some of the allele of the chromosome. The survival strength of an individual is measured using a fitness function which reflects the objectives and constraints of the problem to be solved. After each generation, individuals may undergo culling, or individuals may survive to the next generation (referred to as elitism).

The most popular techniques of evolutionary computing are given below:

- **Genetic Algorithm** Genetic algorithm (GA) is a heuristic approach on the evolutionary ideas of natural selection and genetics. The three basic rules used for creation of next generation are selection of population rule, crossover rule, and mutation rule.
- **Genetic Programming** Genetic programming (GP) is an automated method for creating a working computer program from a high-level problem statement. It starts from a high-level statement of "what needs to be done" and automatically creates a computer program to solve the problem. It is a specialization of the genetic algorithms (GA) where each individual is a computer program. It is a machine learning technique used to optimize a population of computer programs to perform a given computational task.
- **Swarm Intelligence** Swarm Intelligence (SI) is mainly concerned with the design of intelligent multi-agent systems whose inspiration is taken from the collective behavior of social and eusocial insects and other animal populations [7]. The popular algorithms based on SI are Particle Swarm Optimization (PSO) [38, 66], Ant Colony Optimization, Artificial Bee Colony Algorithm, etc.

3.1.4 Artificial Life

Artificial Life are the human-made systems to understand the nature through modelling and simulation. The artificial life models offer the advantage of coding an organism's behavior explicitly as a program, rather than implicitly as the solution to equations that must be integrated. These models are preferred for studying the dynamics of natural evolution.

- **Artificial Immune Systems** As the name suggests, the Artificial Immune System (AIS) is an abstract or artificial component of the natural immune system. The AIS is a powerful information processing and problem-solving paradigm in both the scientific and engineering fields. It possesses nonlinear classification properties along with the biological properties such as self identification, positive and negative selection, clonal selection, etc. The most useful application of AIS is the computer security through detecting viruses and trojans. The other applications are abnormal detection, fault detection, learning and optimization of system.

3.2 Applications of CI

There are several applications of computational intelligence in modeling and designing intelligent systems and solving the real-world problems. The genetic algorithm [20, 25] can be applied to routing optimization in telecommunications networks. The genetic programming can be used in symbolic function identification, empirical discovery, solving systems of equations, concept formation, automatic programming, pattern recognition, game-playing strategies, and neural network design. The application of evolutionary programming [75] is to evolve finite-state machines, optimize a continuous function, and train a neural network (NN). The real-world applications of evolutionary programming are in controller design, robotics, video games, image processing, power systems, scheduling and routing, model selection, design etc. The differential evolution has mostly been applied to optimize the functions defined over continuous-valued landscapes. It can also be applied to train neural networks. The PSO [38] has been used mostly to optimize the functions with continuous-valued parameters. The Artificial Immune Systems (AIS) [55] have been successfully applied to many problem domains. Some of these domains are network intrusion and anomaly detection to data classification models, virus detection, concept learning, data clustering, robotics, pattern recognition and data mining. The AIS has also been applied to initialization of feed-forward neural network weights, initialization of centers of a radial basis function neural network, and optimization of multi-modal functions.

Fig. 2 Combination of three
key terms WSNs, CI and
Security

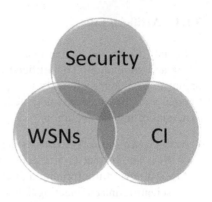

Fig. 2 Combination of three key terms WSNs, CI and Security

4 CI-Based Secure Schemes for WSNs

To study the existing CI-based security schemes for WSNs, we need to consider three
key terms i.e., WSNs, CI, and Security, and their all possible combinations.

From Fig. 2 we can draw all possible combination of schemes for further study
as shown in Table 1. Due the focus on this current study, we will mainly concentrate
on the CI-based secure schemes for WSNs which is the first combination i.e., type
(i) in table. But, as we have already discussed above that this type of schemes are
still under development stage and we have not found more research on this particular
type. So, we will try to discuss some more related schemes of other types [mainly
type (ii), (iv), and (vi)] along with type (i).

Table 1 Various schemes with different combinations

Types	Possible combinations	Representation of possible combinations
i.	CI-based secure schemes for WSNs	$(WSNs + CI + Security)$
ii.	CI-based secure schemes for other than WSNs	$(-WSNs + CI + Security)$
iii.	CI-based schemes without any security issue for other than WSNs	$(-WSNs + CI - Security)$
iv.	CI-based schemes without any security issue for WSNs	$(WSNs + CI - Security)$
v.	Non CI-based security schemes for other than WSNs	$(-WSNs - CI + Security)$
vi.	Non CI-based schemes with security issues for WSNs	$(WSNs - CI + Security)$
vii.	Non CI-based schemes without any security issue for WSNs	$(WSNs - CI - Security)$

4.1 Existing CI-Based Secure Schemes for WSNs

4.1.1 Granular Computing

- **Fuzzy Sets Based Secure Schemes for WSNs** Nghiem and Cho [57] have proposed a multi-hop authentication scheme based on fuzzy logic techniques like fuzzification, fuzzy inference, and defuzzification. They try to address the following problems: (i) wastage of energy during authentication and transmission phases, (ii) key-sharing mechanism among nodes, and (iii) false positive attacks. The claims made in this paper are: (i) by introducing the forwarding and skipping nodes, the wastage of memory is minimized, (ii) a hash-based key sharing method is used for protected key-sharing, and (iii) probabilistic voting-based filtering scheme (PVFS) [43] is used to resist the scheme against the false positive attacks.

The fuzzy logic system used in this scheme have input parameters as REMAINING_ENERGY, HOP_COUNT and FALSE_MAC_COUNT, each having three labels small(S), medium(M) and large(L). The output value, i.e. VERIFICATION_INTERVAL also has three labels {S,M,L}. During fuzzy if-then rule, the three input variables and their labels produce total $27(=3^3)$ rules. The centroid average method have been used for defuzzification to get the crisp value from the fuzzy output facts. From security analysis, it has been observed that this scheme is resilient against the cluster insider attacks, en-route insider attacks, false negative and false positive attacks. Though this scheme provides network security, resists against several attacks and saves up to 13 % of total energy consumption, but it has enough space for further modification in initialization and key assignment mechanism. The hash function based key generation from associated nodes can further be modified to an independent key generation process.

Zou and Liu [86] have proposed a model for evaluating the security of WSNs. They have proposed an interval-valued intuitionistic fuzzy hybrid geometric (IVIFHG) operator to the multiple attribute decision making (MADM) problems for evaluating the network security with interval-valued intuitionistic fuzzy information. In interval-valued intuitionistic fuzzy set (IVIFS) [2], the values of its membership and non-membership functions are intervals, not the exact numbers. The score function (S) and the accuracy function (H) are the difference and the sum of membership and non-membership functions, respectively. It calculates the score $S(\tilde{r}_i)$ of the collective overall interval-valued intuitionistic fuzzy preference values $(\tilde{r}_i)$ to rank all alternative networks systems A_i. The one which has least difference between $S(\tilde{r}_i)$ and $S(\tilde{r}_j)$ is selected. Then it calculates the degrees of accuracy $H(\tilde{r}_i)$ and $H(\tilde{r}_j)$ and ranks the alternatives A_i and A_j accordingly. It ranks all the alternatives A_i and finally selects the best network system in accordance with $S(\tilde{r}_i)$ and $H(\tilde{r}_i)$, where $i = 1, 2, \ldots, m$. This scheme is an innovative one to select the best network, but is not practically implemented in dynamic adaptive environment where the number of sensor nodes are not fixed.

Mishra et al. [54] have proposed an energy efficient packet loss preventive routing protocol based on fuzzy logic. This protocol selects energy-efficient routes to a destination that ensures congestion control, less packet loss, and security. This scheme consists of two phases, route discovery and choice of route. It gives better throughput as compared to the dynamic source routing (DSR) and Ad hoc On Demand Distance Vector Routing (AODV).

Some other fuzzy based secure schemes for WSNs have been proposed by Bolourchi et al. [10], Cheng et al. [13], Choia et al. [14], Fu et al. [21], Huang et al. [24], Lee [41], Li et al. [44], Ren et al. [69], Renubala et al. [71] and Sakthidevi et al. [73].

- **Rough Sets Based Secure Schemes for WSNs** Lin et al. [46] have proposed a scheme for fault diagnosis for WSNs based on rough set theory. Rough set theory deals with the problem having less or incomplete information and can easily resolve the fault diagnosis consuming low energy using datum reduction theory. Even though some redundant attributes have few information, fault of nodes in WSNs can also be accurately diagnosed. This scheme improves the robustness of fault diagnosis of nodes in WSNs with very limited energy consumption.

Some rough set based secure schemes for WSNs have been proposed by Hai-Yang et al. [23], Li et al. [45] and Zhi-Feng et al. [84].

4.1.2   Neuro-Computing

- **Neural Network Based Secure Schemes for WSNs** Kulkarni et al. [36] have proposed a neural network based secure media access control protocol for WSNs. It is a multilayer perceptron (MLP) based media access control (MAC) protocol to secure WSNs. For training the MLP, the backpropagation and particle swarm optimization (PSO) algorithms are used.

The multilayer perceptron (MLP) used in this scheme has three inputs X_1, X_2, X_3 with a constant bias θ (which is set to unity) as shown in Fig. 3. These inputs along with bias are weighted and aggregated in neurons in the hidden layer. The two neurons in hidden layer aggregate the weighted inputs to produce aggregation outputs a_i where $i = 1, 2$ as follows:

$$a_i = \sum_{j=1}^{3} X_j \cdot W_{ij} + \theta \cdot W_{i4} \tag{1}$$

The decision vector d_i which is the set of output of hidden layer is calculated as follows:

$$d_i = \frac{1}{1 - e^{-a_i}} \tag{2}$$

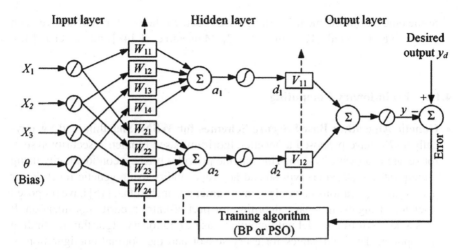

Fig. 3 Structure of multilayer perceptron used in secure MAC [36]

The above outputs then act as inputs to the output layer through a weighted vector $V = \{V_{11}, V_{12}\}$ and produce a final output y as follows:

$$y = d_1 \cdot V_{11} + d_2 \cdot V_{12} \qquad (3)$$

From the above training process the difference between actual output and the desired output is recorded as an error and it is modified until the desired mean square error is achieved. When the output of the MLP exceeds a preset threshold level, the physical layer of node is switched off. This scheme results in saving the power that would have been wasted in retransmission of collided packets.

Some other Neural Network based secure schemes for WSNs have been proposed by Bokareva et al. [9], Kiani et al. [28], Kim et al. [31], Kulakov et al. [34], Kulkarni et al. [37], Padmavathi et al. [59], Ramesh et al. [68] and Ren et al. [70].

- **Reinforcement Learning Based Secure Schemes for WSNs** Usaha and Maneenil [78] have proposed a Reinforcement Learning (RL) to identify the malicious nodes in a network, which is based on the on-policy Monte Carlo (ONMC) method to characterize the reputation values among the nodes and to find a rule for selecting the neighboring nodes based on reputation values which optimizes some performance criterion. Since the nodes have finite buffer to store the arriving packets which are yet to be processed, they follow M/M/1/K *queueing discipline* in which the good nodes have large buffer to receive and forward more packets, but the malicious nodes have smaller buffers resulting in frequent packet drop. The simulation result show that the ONMC scheme gives better packet arrival rate as compared to other existing schemes. The throughput increases up to 71 % for static and 61 % for the dynamic topology.

Some other Reinforcement Learning based secure schemes for WSNs have been proposed by Lee et al. [42], Liu et al. [47], Maneenil et al. [48] and Yau et al. [82].

4.1.3 Evolutionary Computing

- **Genetic Algorithm Based Secure Schemes for WSNs** Shanthini and Swamy-nathan [76] have proposed a genetic algorithm based biometric security system for secure healthcare system. This scheme provides privacy, confidentiality using fingerprint key based encryption and key revocability using genetic function for healthcare applications in wireless environment. Concho et al. [15] have proposed a scheme using evolutionary algorithm for port-of-entry security optimization. It uses a decision-tree model with motive to use evolutionary algorithm for finding the optimal threshold values for every sensor and the optimal configuration of the inspection strategy. The result of probabilistic solution discovery algorithm (PSDA) indicates that it reduces the cost while providing the optimal threshold value. It can further be improved by introducing multi-objective functionality to reduce the cost, time, and to provide the best threshold value for each sensor node. Some Genetic Algorithm based secure schemes for WSNs have been proposed by Lai et al. [40], Shahabadkar et al. [75], Vyas et al. [79] and Zhang et al. [83].
- **Swarm Intelligence Based Secure Schemes for WSNs** Kassabalidis et al. [27] have proposed a particle swarm optimization (PSO) based scheme to identify points on the security border of a power system. To identify the security border, the original feature is tested by selecting the feature set as input to PSO. The PSO uses the neural network for security index (SI) and finally the SI compares it with the desired SI. If the difference is small, the algorithm terminates after providing the point on the border; otherwise, loop continues until the difference comes at satisfactory level or the number of iterations reaches the maximum limit. This technique can be applicable wherever a desired border is required.

Mármol et al. [50] have proposed a bio-inspired technique for providing trust in WSNs. The main aim of this model is to help a node requesting a certain service to the network to find the most trustworthy route leading to a node providing the right requested service. A node can be untrustworthy if it intentionally provides fraudulent or wrong service due to hardware failures or the performance deteriorates. It performs very well when the malicious servers are less than 90% and beyond that it does not provide better chance to select the trustworthy server.

Some Swarm Intelligence based secure schemes have been proposed by Dhurandher et al. [16], Dressler et al. [17], Muraleedharan et al. [56], Periyanayagi et al. [63], Qureshi et al. [67] and Xi et al. [80].

Table 2 Existing CI-based secure schemes for WSNs

CI Paradigms	Research papers	Techniques used	Remarks
Granular computing	Nghiem and Cho [57]	Fuzzy logic techniques like fuzzification, fuzzy inference and defuzzification	Resists the proposed scheme against false positive attacks
	Zou and Liu [86]	Interval-valued intuitionistic fuzzy set	An innovative scheme to choose best network but is not practically implemented in dynamic adaptive environment where the number of sensor node are not fixed
	Mishra et al. [54]	Routing protocol based on fuzzy logic	The proposed scheme gives better throughput than DSR and AODV
	Lin et al. [46]	Rough Sets based fault diagnosis for WSNs	Resolve the problem of fault diagnosis of nodes in WSN more accurately with using limited energy
Neuro-Computing	Kulkarni et al. [36]	Neural Network based secure scheme for WSNs	This multilayer perceptron (MLP) based research helps in extending the lifetime of the WSN. The proposed scheme provides distributed security against the collision attacks. Based on this research an energy model can be built with proper simulation
	Usaha and Maneenil [78]	Reinforcement Learning based secure scheme	On-policy Monte Carlo (ONMC) method along with a reputation scheme select cooperative nodes as well as avoiding malicious nodes with high throughput. This can further gives an idea that reinforcement learning with reputation values can lead to better decision rules for neighboring node selection

Table 2 (continued)

CI Paradigms	Research papers	Techniques used	Remarks
Evolutionary computing	Shanthini and Swamynathan [76]	Genetic Algorithm based biometric security for secure health care system	The fingerprint based cryptographic key is randomized using a genetic operator and provides confidentiality and key revocability for healthcare applications in wireless sensor environments
	Concho et al. [15]	A decision-tree model with use of evolutionary algorithm	Finding the optimal threshold values for every sensor which can further be improved by introducing multi-objective functionality that may reduce the cost and time
	Kassabalidis et al. [27]	Particle Swarm Optimization (PSO) based scheme to identify points on the security border of the power system	This technique can be used further where desire border is required
	Mármol et al. [50]	A bio-inspired technique for providing trust in WSNs	The proposed model performs well only when the malicious servers are less than 90%
Artificial life	Kumar et al. [39]	Artificial Immune Systems based secure schemes for detecting spoofing attacks in WSNs	The scheme is more energy efficient than RSA and ECC based schemes that can be used in future sensor networks

4.1.4 Artificial Life

- **Artificial Immune Systems Based Secure Schemes for WSNs** Kumar et al. [39] have proposed a random key distribution based Artificial Immune System (AIS) for detecting the spoofing attacks in WSNs. This scheme is proved to be secure with detection rate above 90%. The simulation results also prove that it is more energy efficient than the RSA and ECC based schemes and can be used in future sensor networks.

Some Artificial Immune Systems based secure schemes for WSNs have been proposed by Morteza et al. [55], Phogat et al. [65] and Salam et al. [74].

5 Results and Discussions

From reviews and analysis of various existing schemes and Table 2, it is evident that the computational intelligence (CI) based schemes are being used for solving many complex problems of computing. The biologically inspired techniques give us an idea to take steps for finding the solution of complex problem just like nature has been doing for last several years. Many challenges in computer science have also been solved by researchers by applying CI paradigms. The wireless sensor networks (WSNs) is one area where the CI techniques work very well and solve many challenging problems. Security is one of the major issues in any area of computer science. But, very surprisingly, the CI based secure schemes are not so popular and hence have not much been developed so far. From this chapter, we can observe that most of the CI paradigms are not popular for the development of security schemes for WSNs. This systematic study of each and every paradigm of CI can help a researcher to go through all the research works on CI paradigms in one pass.

6 Conclusions and Future Research Directions

In this chapter we have first discussed the WSNs with their types, applications, and security challenges. Then we have presented the CI and its paradigms in brief. Thereafter, we have presented a systematically review of the CI based secure schemes for WSNs. We have analyzed the CI based secure schemes for WSNs. From this study, it is clear that the CI based schemes can help to solve many complex problems. Security is itself a challenging problem. For wireless medium the security has become an essential factor to be considered for any new research or development. Unfortunately, the CI is not a very popular technique in security field. In this chapter, we have tried to present the current state-of-the-art of security related research using CI. With this study, a researcher can proceed further to do research on the CI based security schemes to solve the complex problems of WSNs.

References

1. Agarwal, T.: Wireless Sensor Networks and Their Applications. Accessed on 30 March 2016. http://www.elprocus.com/introduction-to-wireless-sensor-networks-types-and-applications//
2. Atanassov, K.T.: More on intuitionistic fuzzy sets. Fuzzy Sets Syst **33**(1), 37–45 (1989)

3. Bader, S.: Enabling Autonomous Envionmental Measurement Systems with Low-Power Wireless Sensor Networks, Mid Sweden University licentiate thesis, ISSN: 1652-8948, ISBN: 978-91-86694-14-2 (2011)
4. Bargiela, A., Pedrycz, W.: Granular Computing: An Introduction, vol. 717. Springer, Berlin (2012)
5. Bezdek, J.C.: What is computational intelligence?. CONF-9410335– ON: DE95011702; TRN: 95:004731–0002, United States (1994)
6. Biswas, K., Muthukkumarasamy, V., Singh, K.: An encryption scheme using chaotic map and genetic operations for wireless sensor networks. IEEE Sens. J. 15(5), 2801–2809 (2015)
7. Blum, C., Groß, R.: Swarm intelligence in optimization and robotics. In: Springer Handbook of Computational Intelligence, pp. 1291–1309. Springer, Berlin (2015)
8. Bojkovic, Z.S., Bakmaz, B.M., Bakmaz, M.R.: Security issues in wireless sensor networks. Int. J. Commun. 2(1), 106–115 (2008)
9. Bokareva, T., Bulusu, N., Jha, S.: An immunology inspired approach to robust sensor networking. In: Web Abstract of 4th Information Processing in Sensor Networks (IPSN 2005), Los Angeles. pp. 1–2 (2005)
10. Bolourchi, P., Uysal, S.: Forest fire detection in wireless sensor network using fuzzy logic. In: Computational Intelligence, Communication Systems and Networks (CICSyN), 2013 Fifth International Conference on, pp. 83–87. IEEE (2013)
11. Boukerche, A.: Algorithms and Protocols for Wireless Sensor Networks, vol. 62. Wiley, New York (2008)
12. Chauvin, Y., Rumelhart, D.E.: Backpropagation: Theory, Architectures, and Applications, Lawrence Erlbaum Associates, Hillsdale, New Jersey, USA (1995)
13. Cheng, P.C., Rohatgi, P., Keser, C., Karger, P., Wagner, G.M., Reninger, A.S., et al.: Fuzzy multi-level security: an experiment on quantified risk-adaptive access control. In: Security and Privacy, 2007. SP'07. IEEE Symposium on. pp. 222–230. IEEE (2007)
14. Choi, Y., Lee, Y., Won, D.: Security improvement on biometric based authentication scheme for wireless sensor networks using fuzzy extraction. Int. J. Distrib. Sens. Netw. 2016, 1–16 (2016)
15. Concho, A.L., Ramirez-Marquez, J.E.: An evolutionary algorithm for port-of-entry security optimization considering sensor thresholds. Reliab. Eng. Syst. Saf. 95(3), 255–266 (2010)
16. Dhurandher, S.K., Misra, S., Obaidat, M.S., Gupta, N.: An ant colony optimization approach for reputation and quality-of-service-based security in wireless sensor networks. Secur. Commun. Netw. 2(2), 215–224 (2009)
17. Dressler, F., Krüger, B., Fuchs, G., German, R.: Self-organization in sensor networks using bio-inspired mechanisms. In: ARCS Workshops, pp. 139–144 (2005)
18. Du, J., Li, J.: A study of security routing protocol for wireless sensor network. In: Instrumentation, Measurement, Computer, Communication and Control, 2011 First International Conference on, pp. 236–240. IEEE (2011)
19. Engelbrecht, A.P.: Computational Intelligence: An Introduction. Wiley, New York (2007)
20. Ferentinos, K.P., Tsiligiridis, T.A.: Adaptive design optimization of wireless sensor networks using genetic algorithms. Comput. Netw. 51(4), 1031–1051 (2007)
21. Fu, Y., Wu, X.P., Ye, Q., Peng, X.: An approach for information systems security risk assessment on fuzzy set and entropy-weight. Acta Electronica Sinica, 38(7), 1489–1494 (2010)
22. Guo, H., Low, K.S., Nguyen, H.A.: Optimizing the localization of a wireless sensor network in real time based on a low-cost microcontroller. IEEE Trans. Ind. Electron. 58(3), 741–749 (2011)
23. Hai-Yang, Z., Lin, L.: Fault diagnosis of node in wireless sensor network based on the interval-numbers rough neural network. In: Information Management and Engineering (ICIME), 2010 The 2nd IEEE International Conference on, pp. 535–538. IEEE (2010)
24. Huang, X., Sharma, D., Cui, H.: Fuzzy controlling window for elliptic curve cryptography in wireless sensor networks. In: Information Networking (ICOIN), 2012 International Conference on, pp. 312–317. IEEE (2012)

25. Jia, J., Chen, J., Chang, G., Tan, Z.: Energy efficient coverage control in wireless sensor networks based on multi-objective genetic algorithm. Comput. Math. Appl. **57**(11), 1756–1766 (2009)
26. Kaelbling, L.P., Littman, M.L., Moore, A.W.: Reinforcement learning: a survey. J. Artif. Intell. Res. **4**, 237–285 (1996)
27. Kassabalidis, I.N., El-Sharkawi, M., Marks, R.J., Moulin, L.S.: Alves da Silva, A.P., et al.: Dynamic security border identification using enhanced particle swarm optimization. IEEE Trans. Power Syst. **17**(3), 723–729 (2002)
28. Kiani, F., Amiri, E., Zamani, M., Khodadadi, T., Manaf, A.A.: Efficient intelligent energy routing protocol in wireless sensor networks. Int. J. Distrib. Sens. Netw. 1–13 (2015). doi:10.1155/2015/618072
29. Kifayat, K., Merabti, M., Shi, Q., Llewellyn-Jones, D.: Security in wireless sensor networks. In: Handbook of Information and Communication Security, pp. 513–552. Springer, Berlin (2010)
30. Kim, T.K., Seo, H.S.: A trust model using fuzzy logic in wireless sensor network. Int. J. Electr. Comput. Energ. Electron. Commun. Eng. **2**(6), 1051–1054 (2008)
31. Kim, Y.J., Song, S.: The feasibility study of attacker localization in wireless sensor networks. In: Ubiquitous Computing and Multimedia Applications, pp. 180–190. Springer, Berlin (2011)
32. Konar, A.: Computational Intelligence: Principles, Techniques and Applications. Springer, Berlin (2006)
33. Kryszkiewicz, M.: Rough set approach to incomplete information systems. Inf. Sci. **112**(1), 39–49 (1998)
34. Kulakov, A., Davcev, D.: Tracking of unusual events in wireless sensor networks based on artificial neural-networks algorithms. In: Information Technology: Coding and Computing, 2005. ITCC 2005. International Conference on. vol. 2, pp. 534–539. IEEE (2005)
35. Kulkarni, R.V., Förster, A., Venayagamoorthy, G.K.: Computational intelligence in wireless sensor networks: a survey. IEEE Commun. Surv. Tutor. **13**(1), 68–96 (2011)
36. Kulkarni, R.V., Venayagamoorthy, G.K.: Neural network based secure media access control protocol for wireless sensor networks. In: Neural Networks, 2009. IJCNN 2009. International Joint Conference on, pp. 1680–1687. IEEE (2009)
37. Kulkarni, R.V., Venayagamoorthy, G.K., Thakur, A.V., Madria, S.K.: Generalized neuron based secure media access control protocol for wireless sensor networks. In: Computational Intelligence in Multi-Criteria Decision-Making, 2009. mcdm'09. IEEE Symposium on, pp. 16–22. IEEE (2009)
38. Kulkarni, R.V., Venayagamoorthy, G.K.: Particle swarm optimization in wireless-sensor networks: a brief survey. IEEE Trans. Syst. Man Cybern. Part C Appl. Rev. **41**(2), 262–267 (2011)
39. Kumar, E.S., Kusuma, S., Kumar, B.: A random key distribution based artificial immune system for security in clustered wireless sensor networks. In: Electrical, Electronics and Computer Science (SCEECS), 2014 IEEE Students' Conference on, pp. 1–7. IEEE (2014)
40. Lai, C.C., Ting, C.K., Ko, R.S.: An effective genetic algorithm to improve wireless sensor network lifetime for large-scale surveillance applications. In: Evolutionary Computation, 2007. CEC 2007. IEEE Congress on, pp. 3531–3538. IEEE (2007)
41. Lee, H.Y.: Fuzzy-based adaptive countering method against false endorsement insertion attacks in wireless sensor networks. Int. J. Distrib. Sens. Netw. 1–11 (2015). doi:10.1155/2015/484820
42. Lee, M., Ye, X., Johnson, S., Marconett, D., Chaitanya, V., Vemuri, R., Ben Yoo, S.: Cognitive security management with reputation based cooperation schemes in heterogeneous networks. In: Computational Intelligence in Cyber Security, 2009. CICS'09. IEEE Symposium on, pp. 19–23. IEEE (2009)
43. Li, F., Wu, J.: A probabilistic voting-based filtering scheme in wireless sensor networks. In: Proceedings of the 2006 International Conference on Wireless Communications and Mobile Computing, pp. 27–32. ACM (2006)
44. Li, J., Wang, Q., Wang, C., Cao, N., Ren, K., Lou, W.: Fuzzy keyword search over encrypted data in cloud computing. In: INFOCOM, 2010 Proceedings IEEE, pp. 1–5. IEEE (2010)
45. Li, T., Fei, M.: Information fusion in wireless sensor network based on rough set. In: Network Infrastructure and Digital Content, 2009. IC-NIDC 2009. IEEE International Conference on, pp. 129–134. IEEE (2009)

46. Lin, L., Wang, H.j., Dai, C.l.: Fault diagnosis for wireless sensor network's node based on hamming neural network and rough set. In: Robotics, Automation and Mechatronics, 2008 IEEE Conference on, pp. 566–570. IEEE (2008)
47. Liu, Z., Elhanany, I.: Rl-mac: A qos-aware reinforcement learning based mac protocol for wireless sensor networks. In: Networking, Sensing and Control, 2006. ICNSC'06. Proceedings of the 2006 IEEE International Conference on, pp. 768–773. IEEE (2006)
48. Maneenil, K., Usaha, W.: Preventing malicious nodes in ad hoc networks using reinforcement learning. In: Wireless Communication Systems, 2005. 2nd International Symposium on, pp. 289–292. IEEE (2005)
49. Marks II, R.J.: Intelligence: computational versus artificial. IEEE Trans. Neural Netw. 4(5), 737–739 (1993)
50. Mármol, F.G., Pérez, G.M.: Providing trust in wireless sensor networks using a bio-inspired technique. Telecommun. Syst. 46(2), 163–180 (2011)
51. Martins, D., Guyennet, H.: Wireless sensor network attacks and security mechanisms: a short survey. In: Network-Based Information Systems (NBiS), 2010 13th International Conference on, pp. 313–320. IEEE (2010)
52. McCulloch, W.S., Pitts, W.: A logical calculus of the ideas immanent in nervous activity. Bull. Math. Biophys. 5(4), 115–133 (1943)
53. Minsky, M., Papert, S.: Perceptrons, MIT Press, Cambridge, MA (1969)
54. Misra, S., Roy, S., Obaidat, M.S., Mohanta, D.: A fuzzy logic-based energy efficient packet loss preventive routing protocol. In: Performance Evaluation of Computer & Telecommunication Systems, 2009. SPECTS 2009. International Symposium on. vol. 41, pp. 185–192. IEEE (2009)
55. Morteza, J., Hossein, M., Kasra, M., Mohammad, F., Shahaboddin, S.: A Method in Security of Wireless Sensor Network Based on Optimized Artificial Immune System in Multi-Agent Environments. arXiv preprint arXiv:1508.01706 (2015)
56. Muraleedharan, R., Osadciw, L.A.: Jamming attack detection and countermeasures in wireless sensor network using ant system. In: Proceedings of the SPIE 6248, Wireless Sensing and Processing (2006)
57. Nghiem, T.P., Cho, T.H.: A fuzzy-based interleaved multi-hop authentication scheme in wireless sensor networks. J. Parallel Distrib. Comput. 69(5), 441–450 (2009)
58. Padmavathi, G., Shanmugapriya, D.: A Survey of Attacks, Security Mechanisms and Challenges in Wireless Sensor Networks. arXiv preprint arXiv:0909.0576 (2009)
59. Padmavathi, G., Shanmugapriya, D., Kalaivani, M.: Neural network approaches and mspca in vehicle acoustic signal classification using wireless sensor networks. In: Communication Control and Computing Technologies (ICCCCT), 2010 IEEE International Conference on, pp. 372–376. IEEE (2010)
60. Pawlak, Z.: Rough Sets: Theoretical Aspects of Reasoning About Data, vol. 9. Springer, Berlin (2012)
61. Pearl, J.: Probabilistic Reasoning in Intelligent Systems: Networks of Plausible Inference. Morgan Kaufmann, Los Altos, CA (2014)
62. Pedrycz, W.: Shadowed sets: representing and processing fuzzy sets. IEEE Trans. Syst. Man Cybern. Part B Cybern. 28(1), 103–109 (1998)
63. Periyanayagi, S., Sumathy, V.: Swarm based defense technique for denial-of-sleep attacks in wireless sensor networks. Int. Rev. Comput. Softw. (IRECOS) 8(6), 1263–1270 (2013)
64. Perrig, A., Stankovic, J., Wagner, D.: Security in wireless sensor networks. Commun. ACM 47(6), 53–57 (2004)
65. Phogat, S., Gupta, N.: Basics of artificial immune system and its applications. Int. J. Sci. Res. Educ. (IJSRE) 3(5), 3509–3516 (2015)
66. Poli, R., Kennedy, J., Blackwell, T.: Particle swarm optimization. Swarm Intell. 1(1), 33–57 (2007)
67. Qureshi, S., ul Asar, A.: Detection of malicious beacon node based on intelligent water drops algorithm. In: Proceedings on the International Conference on Artificial Intelligence (ICAI). p. 1. The Steering Committee of The World Congress in Computer Science, Computer Engineering and Applied Computing (WorldComp) (2012)

68. Ramesh, M.V., Raj, A.B., Hemalatha, T.: Wireless sensor network security: real-time detection and prevention of attacks. In: Computational Intelligence and Communication Networks (CICN), 2012 Fourth International Conference on, pp. 783–787. IEEE (2012)

69. Ren, Q., Liang, Q.: Fuzzy logic-optimized secure media access control (FSMAC) protocol wireless sensor networks. In: Computational Intelligence for Homeland Security and Personal Safety, 2005. CIHSPS 2005. Proceedings of the 2005 IEEE International Conference on, pp. 37–43. IEEE (2005)

70. Ren, W., Song, J., Ma, Z., Huang, S.: Towards a bio-inspired security framework for mission-critical wireless sensor networks. In: Computational Intelligence and Intelligent Systems, vol. 51, pp. 35–44. Springer, Berlin (2009)

71. Renubala, S., Dhanalakshmi, K.S.: Trust based secure routing protocol using fuzzy logic in wireless sensor networks. In: Computational Intelligence and Computing Research (ICCIC), 2014 IEEE International Conference on, pp. 1–5. IEEE (2014)

72. Rosenblatt, F.: The perceptron: a probabilistic model for information storage and organization in the brain. Psychol. Rev. 65(6), 386–408 (1958)

73. Sakthidevi, I., Srievidhyajanani, E.: Secured fuzzy based routing framework for dynamic wireless sensor networks. In: Circuits, Power and Computing Technologies (ICCPCT), 2013 International Conference on, pp. 1041–1046. IEEE (2013)

74. Salam, A., Nadeem, A., Ahsan, K., Sarim, M., Rizwan, K.: A class based qos model for wireless body area sensor networks. Res. J. Recent Sci. ISSN 2277:2502

75. Shahabadkar, R., Pujeri, R.V.: Secure multimedia transmission in p2p using recurence relation and evolutionary algorithm. In: Security in Computing and Communications, vol. 377, pp. 281–292. Springer, Berlin (2013)

76. Shanthini, B., Swamynathan, S.: Genetic-based biometric security system for wireless sensor-based health care systems. In: Recent Advances in Computing and Software Systems (RACSS), 2012 International Conference on, pp. 180–184. IEEE (2012)

77. Sutton, R.S., Barto, A.G.: Reinforcement Learning: An Introduction, vol. 1. MIT Press, Cambridge (1998)

78. Usaha, W., Maneenil, K.: Identifying malicious nodes in mobile ad hoc networks using a reputation scheme based on reinforcement learning. In: TENCON 2006. 2006 IEEE Region 10 Conference, pp. 1–4. IEEE (2006)

79. Vyas, N., Shah, R.: Intelligent and efficient cluster based secure routing scheme for wireless sensor network using genetic algorithm. Int. J. Digit. Appl. Contemp. Res. 2, 1–7 (2014)

80. Xi, O., Jianyi, Z., Zhe, G., Qi, L.: A reputation-based ant secure routing protocol of wireless sensor networks. Int. J. Adv. Comput. Technol. 4(9), 9–18 (2012)

81. Xia, F., Zhao, W., Sun, Y., Tian, Y.C.: Fuzzy logic control based qos management in wireless sensor/actuator networks. Sensors 7(12), 3179–3191 (2007)

82. Yau, K.L.A., Komisarczuk, P., Teal, P.D.: Reinforcement learning for context awareness and intelligence in wireless networks: review, new features and open issues. J. Netw. Comput. Appl. 35(1), 253–267 (2012)

83. Zhang, J., Lao, Y., Wen, L.: A genetic algorithm approach for security authentication in the wireless sensor networks. In: Wireless Communications, Networking and Mobile Computing, 2007. WiCom 2007. International Conference on, pp. 2259–2263. IEEE (2007)

84. Zhi-Feng, D., Yuan-Xiang, L., Guo-Liang, H., Ya-La, T., Xian-Jun, S.: Uncertain data management for wireless sensor networks using rough set theory. In: Wireless Communications, Networking and Mobile Computing, 2006. WiCOM 2006. International Conference on, pp. 1–5. IEEE (2006)

85. Zhou, J., Pedrycz, W., Miao, D.: Shadowed sets in the characterization of rough-fuzzy clustering. Pattern Recognit. 44(8), 1738–1749 (2011)

86. Zou, P., Liu, Y.: Model for evaluating the security of wireless sensor network in interval-valued intuitionistic fuzzy environment. Int. J. Adv. Comput. Technol. 4(4), 254–260 (2012)

Efficient Anomaly Detection System for Video Surveillance Application in WVSN with Particle Swarm Optimization

S. Radha, S. Aasha Nandhini and R. Hemalatha

Abstract Wireless sensor networks consist of several tiny low cost sensor nodes that are deployed for many applications such as military, civil, industrial, healthcare, home automation, etc. Recent technological developments have enabled the use of wireless visual sensor networks (WVSNs) for sensitive applications such as video surveillance and monitoring applications. Limited memory, energy and bandwidth are the major constraints in WVSN that can be simplified by the use of compressed sensing (CS), which asserts that sparse signals can be reconstructed from very few measurements. CS a computational intelligence solution is about acquiring and recovering the signal in the most efficient manner possible using incoherent projection basis. In the case of video surveillance applications, the entire video may not be useful hence, with the help of efficient algorithms the presence of the anomalies can be detected and transmitted to help user at the monitoring site to take necessary action. In this chapter, particle swarm optimization (PSO) based efficient anomaly detection system (EADS) is proposed which will detect the presence of anomalies and transmit the required measurements via TelosB nodes to the network operator. This system adopts the concept of CS to obtain the compressive measurements so that the object detection algorithm can be applied to the measurements rather than samples. PSO is employed for optimizing compressive measurements while a mean based measurement differencing approach is used for detecting the object. This proposed efficient system has the intelligence of detecting targets with fewer measurements and transmit the required compressive measurements for reconstruction with less energy, thereby increasing the network lifetime. PSO is used to optimize the transmission distance with minimum number of hops towards destination, to achieve reduced energy consumption. However, the lifetime of the network is still bounded by batteries, the sole source of energy in WVSNs. Alternative energy utilization can be effectively included to recharge the batteries on-board and extend the lifetime of

S. Radha (✉) · S. Aasha Nandhini · R. Hemalatha
Department of ECE, SSN College of Engineering, Kalavakkam, Chennai 603 110, India
e-mail: radhas@ssn.edu.in

S. Aasha Nandhini
e-mail: aashanandhinis@ssn.edu.in

R. Hemalatha
e-mail: hemalathar@ssn.edu.in

© Springer International Publishing AG 2017
A. Abraham et al. (eds.), *Computational Intelligence in Wireless Sensor Networks*,
Studies in Computational Intelligence SCI 676, DOI 10.1007/978-3-319-47715-2_7

153

the network. Solar energy harvesting forms an effective resource due to its ambient presence. Hence, solar energy harvester is incorporated in the proposed EADS to extend its lifetime.

Keywords WVSN · Compressed sensing · PSO · Solar harvester · Network lifetime · Video surveillance

1 Introduction

A video surveillance system consists of cameras that can monitor secured areas and transmit the captured video to a monitoring site where the network operator analyses the video on occurrence of an event such as assault or robbery, etc. Most of the time the video captured by the surveillance camera keeps unvarying and thereby transmitting the entire video consumes more energy and bandwidth. An efficient video surveillance system should have the ability to detect the dubious activity with less energy and bandwidth by transmitting the data that are pertinent for surveillance.

In a wireless video surveillance system the captured video is pre-processed and compressed before transmission, and analyzed at the receiving end [1]. Video Surveillance over wireless sensor networks (WSNs) has been widely adopted in various fields such as traffic monitoring, healthcare, public safety, environmental monitoring and anomaly detection. However, the transmission process at each sensor node is still a challenging job for real time video surveillance applications, as it deals with voluminous video data. The process of video analysis can be done at the transmitter or receiver. However, in this work the analysis is performed at the transmitter side to detect the anomaly. The remote control unit at the receiver end can retrieve the information and take necessary action based on such information. The surveillance video with static background is best suited for object detection applications and the commonly used technique is background subtraction. For obtaining the background modeled image, the background must be dynamically updated based on different background models. Object detection and object tracking are the two commonly studied applications of the advanced video surveillance system. In order to process the video in a resource constrained environment like WSN, a promising technique called compressed sensing can be exploited thereby reducing the energy, complexity and bandwidth.

Compressive sensing is an emerging field that reconstructs the original signal from small number of measurements using sub-Nyquist sampling rates [2, 3]. The CS theory shows that a signal can be reconstructed from a small set of random projections, provided the signal is sparse in some basis, e.g., wavelets. CS has been widely used for many resource constraint applications [4–7]. Compressive measurements require less bandwidth and complexity compared to transform coding of the raw data. Once the signal is reduced to few compressive measurements, background subtraction can be carried out for obtaining the foreground measurements. In conventional background subtraction, the subtraction is performed on each pixel resulting in higher computational complexity, whereas in CS based background subtraction, the process is carried out on compressive measurements. CS based background subtraction can also be performed by applying CS directly to the differenced image. The main idea is that the background subtracted images can be represented sparsely in the spatial image domain. Hence, the CS reconstruction theory should be applicable for direct recovery of the foreground. The CS measurements play a major role in determining detection accuracy and transmission energy, hence it is necessary to find the optimal measurements that yields higher detection accuracy with a less energy requirement. Computational intelligence (CI) solution can be used for optimizing the measurements and routing distance for achievement of better detection accuracy and optimum shortest distance towards the destination respectively.

CI is one of the rapidly growing fields for many years attracting a large number of researchers and practitioners working in the area of neural networks, fuzzy logic, evolutionary computing and swarm intelligence. There are many successful applications of CI such as image processing or retrieval, audio processing and text processing [8]. Multimedia communication needs advanced and efficient computational methodologies for dealing with the huge volume of data generated by these applications.

Researchers have used potential CI methods for overcoming the challenges in WSN [9]. There are different CI based solutions for WSN such as evolutionary algorithm for network design, fuzzy logic for network deployment, swarm intelligence for localization and neural networks for security [9]. In [10] the authors have integrated PSO as a robust solution to sensor networks to produce low energy sensor nodes. A PSO based simulator called as PSO SIMSSENS to find the optimal path of the sink node is also proposed. Sequential PSO (S-PSO), a modification of the PSO is used for reducing the dimensionality issues in distributed sonar sensor placement. S-PSO uses fewer particles to solve the low dimensional problem thereby reducing the computational complexity [11]. Available literature shows that PSO has higher flexibility and optimality with less complexity and memory requirements. This makes it ideally suited for addressing energy related issues in WSN compared to other CI solutions. Hence PSO is adopted in this work as it provides optimization solutions to WSN which in turn solves the energy, memory and bandwidth problems in an efficient way.

The objective of this work is to design an efficient anomaly detection system that can detect the object in the video and transmit the required foreground measurements for reconstruction and further analysis. This system is developed targeting the video surveillance application in WVSN by adopting CI solutions. PSO is employed for the optimization of the measurements and distance towards destination to improve the network lifetime. CS is applied to the video frames for obtaining the compressive measurements, the number of measurements is decided with the help of PSO and a mean measurement differencing approach is used to obtain the differenced measurements. The foreground measurements are extracted from the differenced measurements using a threshold strategy. PSO is employed for optimization of the measurements in order to achieve maximum detection accuracy with less energy. Moreover, the transmission distance is also optimized using PSO to enhance the lifetime as the energy consumption and lifetime is mainly dependent on the transmission energy.

Lifetime is bounded by the batteries-the sole source of energy, despite reduction in energy consumption. It is also hard to replace or recharge nodes battery once they are deployed. Energy harvesting can be used for achieving lasting operational lifetime. Solar energy is a natural resource available in abundance. The power conversion efficiency of solar energy harvesters mainly depends on the variation of the maximum power point of the photovoltaic panel due to the temperature or irradiance level change in the environment. Hence, it is essential to use a maximum power point tracker (MPPT) [12] and a suitable converter to track the power changes and extract maximum possible energy from the sun. Though there are several PSO based MPPT's available in literature, they are widely used for partially shaded conditions, hence in this work normal MPPT is used for power tracking [13, 14]. The harvested energy forms the additional resource and helps in improving the entire system lifetime.

The rest of the chapter is organized as follows: Sect. 2 provides an overview of the system model, Sect. 3 explains in detail about the anomaly detection framework for WVSN with solar harvester and the proposed PSO based object detection approach. Section 4 provides the detailed description of the designed solar harvester, Sect. 5 provides the performance evaluation of the entire framework with and without solar harvester. Section 6 gives the conclusion and scope for future work.

2   System Model

Consider an indoor surveillance scenario as shown in Fig. 1 where a network is deployed using wireless sensor nodes for transmission of the information obtained from the EADS system to the network operator. For efficient processing of the video and transmitting the required information with less complexity and energy, CS technique is adopted. PSO is used to optimize the CS measurements and a simple mean measurement differencing approach is proposed for this system for detection of the objects using optimized compressive measurements. The optimized routing path is also obtained by using PSO, to reduce the communication energy. This efficient system is implemented in the camera motes and the foreground measurements are transmitted through relay nodes (via the optimal path) to the network operator. This system mainly targets the indoor surveillance applications as sensor nodes have limited energy. This system can be used to detect the presence of an intruder in a highly secured indoor environment. The camera node must be associated with a passive infrared sensor which can trigger the camera when there is a movement. The captured video is reduced to a low dimensional signal by exploiting the CS technique and then the object detection algorithm is used for extracting the foreground measurements. These measurements are alone transmitted via solar energy harvester equipped relay nodes. At the monitoring site the network operator will reconstruct the object using an efficient CS recovery algorithm.

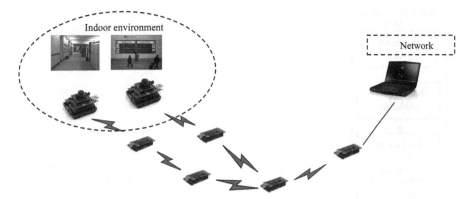

Fig. 1 Indoor surveillance system

3 Efficient Anomaly Detection System

In this framework, EADS system is used for extraction of the foreground measurements and transmits it via TelosB nodes [15]. The mean based differencing approach performed on the compressive measurements reduces the complexity as subtraction is performed on the measurements rather than on all the pixels. PSO is adopted for the optimization of the compressed measurements such that the proposed approach yields better detection accuracy with less energy. The routing path towards the destination is also optimized using PSO to consume lesser energy for transmitting the compressive measurements. A solar harvester is incorporated in the TelosB node for ensuring further increase in the lifetime of the nodes. The block diagram of the PSO based anomaly detection framework is shown in Fig. 2. The captured video is given as input to the background modeling block in which the background is modelled for further processing. The PSO based mean measurement differencing approach yields optimized differenced measurements from which the foreground measurements are extracted using a thresholding strategy. A CS recovery algorithm is used for reconstruction of the compressed measurements. The processes involved are explained in detail below.

3.1 Background Modelling

The frames of size $N \times N$ are extracted from the video and the background is modelled using a running average model [16] which can be efficiently used for sensor networks as it has low computational complexity [16].

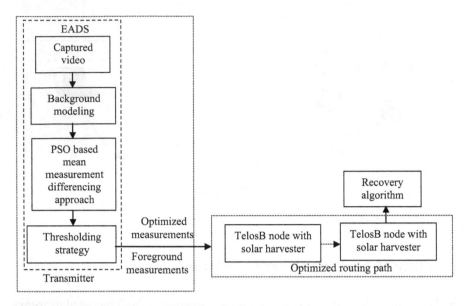

Fig. 2 Block diagram of proposed PSO based object detection approach

The background is modelled as shown in Eq. (1).

$$B_{t+1}(x, y) = \beta\, B_t(x, y) + \alpha\, C_t(x, y) \tag{1}$$

where α is the learning rate, $\beta = 1 - \alpha$, B_t is the initial background frame at time t and C_t is the current input video frame at time t [16]. The first frame is considered as the background frame and B_{t+1} is used as the updated background frame at time $t+1$.

3.2 Compressed Sensing

CS is a signal processing technique which asserts that the signal can be reconstructed from a far few measurements (M) rather than N samples. CS can be applied only to sparse signals and all natural signals are compressible in some form or the other. The signal is said to be compressible in any transform domain such as DCT, DWT etc. The background frame B_t and current frame C_t in sparse domain are represented as in Eqs. (2) and (3)

$$B_t = \psi s_b \tag{2}$$
$$C_t = \psi s_c \tag{3}$$

The measurement vector of the background frame and current frame are obtained by applying the measurement matrix to the input frame as represented in Eqs. (4) and (5)

$$y_b = \Phi B_t \qquad (4)$$
$$y_c = \Phi C_t \qquad (5)$$

where y_b and y_c represents the measurement vector of size $M \times 1$, s_b and s_c represents the sparse vector of background frame and current frame respectively, Φ denotes the measurement matrix of size $M \times N$, Ψ represents the basis matrix, B_t and C_t represent the background frame and current frame respectively [17]. Subsequent to sparsification of the frames, an efficient measurement matrix is applied for obtaining the measurements. The minimum number of measurements 'M' required for perfect reconstruction depends on the sparsity level.

In order to achieve perfect recovery the measurement matrix must satisfy the incoherent property and restricted isometry property. There are many CS recovery algorithms such as basis pursuit [18], orthogonal matching pursuit [19] and iterative algorithms [20–22] that aid in the perfect recovery of the signal. However, in this work Orthogonal Matching Pursuit (OMP) algorithm is utilized. It is a fast and inexpensive algorithm that constructs the estimated sparse vector using iteration process and the number of iterations depends on the sparsity level of the signal [19].

3.3 Particle Swarm Optimization

PSO [23, 24] is a population based stochastic optimization technique developed by Edward and Kennedy in 1995. It is based on bird flocking or fish schooling. PSO shares many similarities with Genetic Algorithms (GA), however PSO has no evolution operators such as crossover and mutation. Initially the PSO system is set with a population of random solutions and converges towards the optima by updating generations. In PSO, the particles move towards the current optimum particles for achievement of the best local and global solution which is explained in detail in Algorithm 1. The PSO is very easy to implement and the parameters to be adjusted are less, which requires reduced computation time and memory [25]. PSO has been successfully applied in many areas: function optimization, artificial neural network training, fuzzy system control and other areas where GA can be applied [26]. The basic steps of the PSO algorithm [27, 28] are given in Algorithm 1.

Initialization:
      (i)Swarm size
      (ii)Velocity and position of each particle
Procedure:
    Step1: For each particle calculate fitness value for each particle. If the fitness value is better than the best fitness   value (Pbest). Set current value as the new Pbest.
    Step 2: Choose the particle with best fitness value of all the particles as Gbest.
    Step 3: For each particle calculate the particle velocity

$$V_i^{(t+1)} = w * V_i^{(t)} + c_1 * rand1() * (Pbest - X_i^{(t)}) + c_2 * rand2() * (Gbest - X_i^{(t)})$$

    Step 4: Update the particle position

$$X_i^{(t+1)} = X_i^{(t)} + V_i^{(t+1)}$$

      Where c_1, c_2 are the scaling coefficients, $X_i^{(t)}$ and $V_i^{(t)}$ are the position and velocity of particle I in t^{th} iteration.
      Until maximum iterations or minimum error is attained

The PSO is used for optimizing the minimum number of compressive measurements and the distance of the routing path, which will improve the lifetime of relay nodes by reducing the communication significantly. The number of foreground measurements required to be transmitted for detection of objects is based on the PSO and adaptive threshold.

3.4   Mean Measurement Differencing Approach

This approach is applied on the optimized compressive measurements of the background frame and current frame. Initially the mean of the background measurements is computed and subtracted from the measurements of the current frame resulting in CS differenced measurements. An adaptive threshold is designed based on the standard deviation of the differenced measurements to extract the required foreground measurements. The mean of the background measurements are computed using Eq. (6).

$$\mu_B = \frac{1}{M_1} \sum_{i=1}^{M_1} y_b(i) \tag{6}$$

where M_1 represents the number of measurements of a block and y_b represents the measurements of the background frame. The differenced measurements 'y_d' are obtained using Eq. (7)

$$y_d = y_c - \mu_B \times U \tag{7}$$

where y_c represents the current frame measurements, U represents an $M_1 \times 1$ matrix with unit entries and μ_B represents the mean of the background measurements.

The threshold (Z) is designed based on the standard deviation of the differenced measurements as shown in Eq. (10).

$$\mu = \frac{1}{M_1} \sum_{i=1}^{M_1} y_d(i) \tag{8}$$

$$\sigma_d = \sqrt{\frac{1}{M_1} \sum_{i=1}^{M_1} (y_d(i) - \mu)^2} \tag{9}$$

$$Z = \sigma_d \tag{10}$$

where μ and σ_d represents the mean and standard deviation of the differenced measurements respectively. The threshold is computed for each block using Eq. (10) and gets adapted based on the difference measurements. The differenced measurements are compared with the adaptive threshold, if the measurements exceed the threshold, they are then transmitted for reconstruction else the measurements are dropped. These transmitted measurements denoted as 'y_f' are called as foreground measurements, with which the object is reconstructed. In this system the foreground measurements are alone transmitted for detection of objects. These foreground measurements are extracted with the help of an adaptive threshold. In order to minimize the foreground measurements, the number of measurements must be minimized at the CS process. The minimum number of measurements computed by the CS process is optimized using PSO such that the object detection algorithm yields maximum detection accuracy with less number of measurements. The objective function that minimizes the measurements is given in Eq. (11)

$$f = \min(y_f) \tag{11}$$

where f represents the optimized foreground measurements and y_f represents the extracted foreground measurements. PSO provides the best minimum measurements for which the detection accuracy is higher resulting in reduced transmission energy. The OMP algorithm is used at the receiver side for reconstruction of the foreground object.

It is sufficient to transmit the foreground measurements extracted from the framework to detect the object. Hence the communication energy is reduced to a great extent, thereby increasing the lifetime of the network. The optimized foreground measurements have to be transmitted efficiently to the destination. The routing path and the energy resource form the major metrics that affects the system lifetime. Hence, PSO is used for optimization of the routing path and solar harvester is used to boost up the energy resource. The energy harvester designed is explained in the following section.

4 Solar Energy Harvester

A simple photovoltaic energy harvesting (PVEH) system is designed as shown in Fig. 3. It consists of the WSN mote as the load.

The energy generated by the solar PV panel is efficiently stored in the battery with the boost converter and the MPPT in action. The duty cycle of the converter is varied using the direct control method based on the output from the MPPT. Compensator circuit is included to ensure stability in the system. As the battery supplies power to the load, to get a voltage level (2.7–3.3 V) suitable for the mote operation, buck regulator is used. Battery charge control is included to avoid over charging and draining of the battery.

4.1 PV Panel Modeling

The Blue Solar SL8585mm PV panel is used for the MATLAB simulation model. The panel provides 950 mW of nominal maximum power. It is of monocrystalline type and well suited to the requirement range of WSN. The simple one diode model [29] is used for modeling the PV panel in simulation.

The voltage–current and voltage–power characteristics of the PV panel at $T = 25\,^\circ C$ and varying irradiance levels are shown in Fig. 4. The V–I and voltage–power characteristics of the PV panel at $G = 1000\,W/m^2$ and varying temperature levels are shown in Fig. 5. Irradiance variation is considered upto $G = 1000\,W/m^2$ as per the standard test conditions. From the Figs. 4 and 5, it is evident that the irradiance change introduces a considerable change in the peak power, whereas the change in temperature does not produce any drastic variation in the peak power. Moreover in Indian scenario the temperature profile during a day does not experience a wide variation, excluding the exceptions on a few rare occasions. Hence the temperature is kept constant for further analysis. However when temperature varying locations are chosen, the effect of temperature must also be analyzed and included.

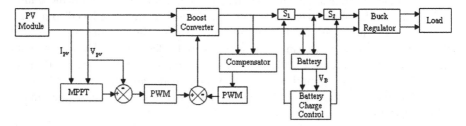

Fig. 3 Block diagram of the proposed PV energy harvesting system

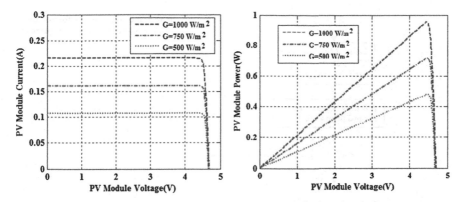

Fig. 4 Simulated V–I and Voltage Power characteristics of the PV panel for varying irradiance at T = 25 °C

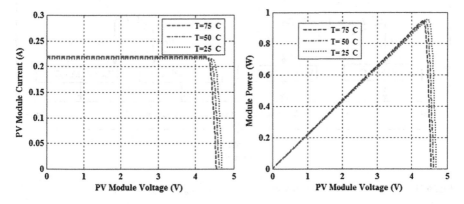

Fig. 5 Simulated V–I and Voltage Power characteristics of the PV panel for varying temperature at G = 1000 W/m^2

4.2 MPPT Converter

Boost converter is used as MPPT. The converter is chosen on the basis of the required battery voltage (4.8–5.6 V) and the available PV voltage level (Vm = 4.5 V). The inductor and capacitor values are designed by considering the operation of converter in continuous conduction mode (CCM). The values of L and C are calculated with voltage and current ripple of 5 %. The ripple percentage is calculated with respect to the expected steady state DC component. The obtained values are L ≥ 107 mH and C ≥ 123.29 μF with switching frequency, fs = 10 kHz. Hence to ensure CCM, L = 200 mH and C = 200 μF have been used in simulation.

MPPT controller is used for matching the source resistance to load resistance as seen by the PV panel (source) to extract maximum power. The duty cycle is set to its optimal value corresponding to the optimal operating point (Vm, Im) using MPPT.

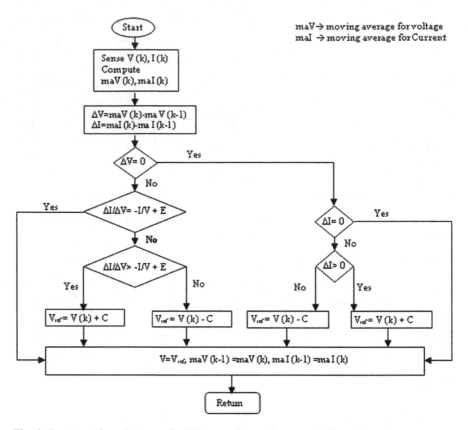

Fig. 6 Incremental conductance algorithm over the moving averaged input

A simple moving averaged incremental conductance (MAIC) algorithm has been employed as shown in Fig. 6. When there is a sudden change in the environment or the climatic condition around the PVEH system over a short duration of time, there is a chance of erroneous result in the maximum power point. A third order moving average filter is included for overcoming this problem. It calculates the average of consecutive three samples of current and voltage of the PV panel. The moving averages are used to calculate the differential voltage and current values, from which the conductance value will be calculated and based on that the panel voltage, will be either increased or decreased. Marginal error (E) of 0.002 is included in this work. Panel voltage determines the duty cycle of the converter.

Moving average filter is implemented in recursive fashion using Eqs. (12) and (13) as it has faster implementation, reduced complexity (one addition and one subtraction after the calculation of the first average) and integer based implementation. The values of p and q are taken as 1 and 2 respectively.

$$maV(i) = maV(i-1) + V(i+p) - V(i-q) \tag{12}$$

$$maI(i) = maI(i-1) + I(i+p) - I(i-q) \tag{13}$$

NiMH battery is used in the PVEH system. Dynamic modeling of the battery is performed in MATLAB by considering both the charging and discharging states [30]. The battery used has a nominal voltage of 4.8 V and fully charged voltage of 5.62 V. The battery state of charge is controlled by two switches and a charge control algorithm, providing series charge regulation of the PVEH system. Whenever the battery terminal voltage is higher than the upper limit of the charging cycle (80 % State of Charge) the switch S1 in Fig. 3 is opened and prevents the overcharging of the battery and increases its lifetime. If the battery voltage is less than the lower critical limit (20 % State of Charge) and there is no power available from the solar panel the switch S2 in Fig. 3 is opened to avoid the deep discharge of the battery. As the battery full charge voltage is 5.62 V, to make it suitable for the operating range of the mote (2.7–3.3 V) step down converter is used after the battery. The duty cycle of the buck converter is adjusted using the voltage mode controller to set the output voltage to approximately 2.7 V. The output voltage can also be fixed for any range between (2.7–3.3 V).

4.3   Simulation Results of PVEH

The simulation schematic of the entire PVEH system is shown in Fig. 7. The PVEH system has been tested with the mote of interest under two different scenarios. In the first test scenario the irradiance level (G) is kept as $1000 \, W/m^2$ and the temperature (T) is taken as 25 °C for the entire simulation duration. Mote is assumed to be in sleep state for the first half of the duration and in active state for the next half. The initial state of charge of the battery is taken as 50 %. Second test scenario has the parameters as $G = 500 \, W/m^2$ for the first 0.5 s and $G = 1000 \, W/m^2$ during the next 0.5 s, T = 25 °C and the mote state is assumed to have a duty cycle of 1 % as most

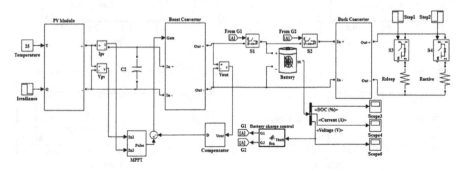

Fig. 7 Schematic of the PVEH system

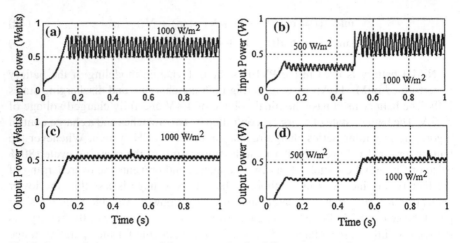

Fig. 8 Input and output power of the converter for the different test scenarios

of the WSN applications have a duty cycle <1%. The input and output power of the converter is observed under these conditions and shown in Fig. 8. The output tracks the corresponding changes in the input, the output and the input power are approximately the same due to the usage of MAIC algorithm.

During the initial 0.5 s the mote is in sleep state, consuming nearly (43–45) μW. During the last 0.5 s the mote is in active state consuming nearly 95–98 mW. Similarly the mote current consumption in sleep state is nearly 16 μA and during active state it consumes nearly 33–36 mA. The power consumption of the mote is as shown in the Fig. 9. The mote power consumption also varies according to changes in the duty cycle.

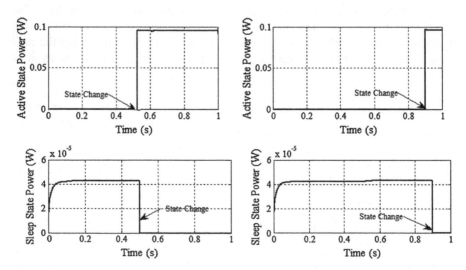

Fig. 9 Power consumption of the mote

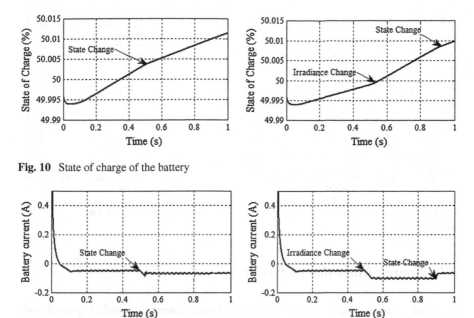

Fig. 10 State of charge of the battery

Fig. 11 Battery current

The battery State of Charge (SOC) in both scenarios is shown in Fig. 10.

It shows that the linear increase in SOC till 0.5 s, then as the mote becomes active, current consumption increases and the battery current reduces accordingly. Hence, the rate of change in SOC decreases after 0.5 s. In Fig. 9 SOC increases linearly upto 0.5 s, at 0.5 s irradiance level change occurs so the current entering the battery is increased. Hence, the rate of increase of SOC also increases. After 0.9 s the mote becomes active so the rate of increase in the SOC decreases. The corresponding change in the battery current is shown in Fig. 10.

The comparison of the output power obtained, with the conventional direct coupling method and the PVEH method is carried out and the results are shown in Fig. 11.

The analysis is carried out under different illumination conditions. The results shown in Fig. 12 are for the irradiance levels 1000 and 800 W/m² at 25 °C respectively. Direct coupling of the mote with solar panel yields very less amount of power conversion due to impedance mismatch. The proposed PVEH produces 85 % of the theoretical maximum power with 87 % efficiency converter on an average. PVEH has a better energy conversion efficiency under varying irradiance and temperature too.

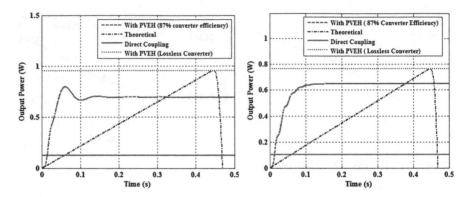

Fig. 12 Comparative analysis of the output maximum power harvested at 1000 W/m^2, 25 °C and 800 W/m^2, 25 °C.

4.4   Energy Harvested by PVEH

The energy harvestable using PVEH in a day is calculated by considering the illumination variations in Chennai region provided by National Renewable Energy Laboratory (NREL) in the renewable resource data centre [31]. Based on the power conversion efficiency of the PVEH and the sample irradiance values as provided in Table 1, energy harvested in a single day is calculated as 186.40 J/h. After including a conversion loss of 20 % the energy obtained is found to be 149.19 J/h. Hence energy harvested per day is calculated as 1193.536 J.

Table 1 Irradiance values for a sample day in Chennai region

Sample irradiance	
Time	Measured irradiance (W/m^2)
7.00 a.m.	1
8.00 a.m.	165
9.00 a.m.	454
10.00 a.m.	702
11.00 a.m.	889
12.00 noon	1004
1.00 p.m.	1037
2.00 p.m.	987
3.00 p.m.	858
4.00 p.m.	657
5.00 p.m.	400
6.00 p.m.	110

With the efficient design of the energy harvester, it is evident that the harvester is capable of harvesting 1193.5636 J/day on an average. This additional energy can be added with the battery resource to calculate the lifetime.

5 Performance Evaluation

The performance of the anomaly detection framework and the impact of the EADS in WSN are evaluated in terms of detection accuracy, communication energy and network lifetime. A network of sensor nodes was formed using TelosB nodes as the relay nodes and PC was used at the source as well as the destination. The PC at the transmitter side acts as the source node in which the EADS was implemented for detection of the object from the captured video. The foreground measurements extracted were transmitted via TelosB nodes to the destination PC where the object is reconstructed using OMP algorithm. The source PC implements the PSO along with the object detection algorithm for optimization of the measurements and distance (with reduced hops) towards destination. The measurements are transmitted in a multihop manner through the sensor nodes. The object detection algorithm and PSO are implemented in MATLAB software in PC and the measurements are transmitted in real time under ContikiOS platform [32]. The EADS is analyzed for conventional CS based background subtraction and proposed mean measurement differencing approaches.

The proposed framework has been tested using two videos, namely, corridor sequence which is captured and lena walk sequence taken from database [33]. Two frames from each sequence are shown to demonstrate the effect of the framework. The frames of size 288×352 are extracted from the video and are divided into blocks of size 8×8. CS is applied to each block to obtain the compressive measurements. A simple and novel matrix called hybrid matrix is used for the CS process using a combination of multiple matrices. Hybrid matrix which is designed by augmenting toeplitz matrix with entries $(-1, +1)$ and binary matrix with entries $(0, 1)$ reduces the storage and energy complexity while maintaining an acceptable range of PSNR [5]. The optimized measurements that maximize the detection accuracy are computed using PSO. The values of the parameters employed in PSO are provided in Table 2.

In conventional CS based background subtraction, the compressive measurements of the background frame and current frame are subtracted for obtaining the

Parameters	PSO value
Population size	12
Number of generations	10
C1	2
C2	2

Table 2 Parameters employed in PSO

Table 3 Simulation parameter table

Frame resolution: 288×352
Block size N: 8
K_1: 5 sparsity level
t $= 1/250$ s for transmit/receive
I_t: 19.8 mA for transmission
I_r: 23 mA for reception
V: 3 V

differenced measurements. The background subtraction process is represented as $y_d = y_c - y_b$, where y_d, y_c and y_b represents the differenced measurements, current frame measurements and background measurements respectively [34]. These differenced measurements are transmitted to the monitoring site. In the proposed approach the differenced measurements and foreground measurements are obtained using Eqs. (6) and (9). The object is reconstructed from the foreground measurements by employing OMP algorithm which is a fast and less expensive algorithm. The proposed object detection approach is compared with the CS based object detection approach in terms of detection accuracy and communication energy. The parameters for simulation are provided in Table 3.

Figure 13a, b shows the input frames, ground truths for corridor sequence and lena walk sequence respectively where the ground truth is generated manually. Figures 14 and 15 shows the reconstructed frames using CS based background subtraction and proposed object detection approach for corridor sequence and lena walk sequence respectively.

(a)

(b)

Fig. 13 Original frames and its corresponding ground truth of Video sequence 1. **a** Corridor sequence 1 and, **b** lena walk

Fig. 14 Object detected using **a** CS based background subtraction and **b** proposed MMD approach with threshold strategy for corridor video sequence

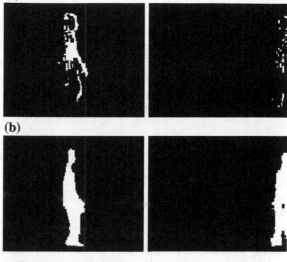

Fig. 15 Object detected using **a** CS based background subtraction and **b** proposed MMD approach with threshold strategy for lena walk video sequence

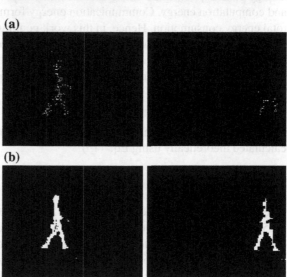

From Figs. 14 and 15 it is clearly seen that the proposed object detection algorithm achieves better detection accuracy compared with the CS based background subtraction algorithm.

5.1 Detection Accuracy

The F_1 score metric is used to evaluate the detection accuracy of the object detection algorithm which is given in Eq. (14). F_1 score is defined as the weighted harmonic mean of recall and precision where recall represents the fraction of the detected foreground pixels to the total number of foreground pixels in the ground truth and precision represents the fraction of the correctly detected foreground pixels to the total number of foreground pixels detected [35].

$$F_1 \text{ score} = 2 * \frac{\text{Recall} * \text{Precision}}{\text{Recall} + \text{Precision}} \tag{14}$$

5.2 Energy Analysis

Energy consumption in a WSN node has two components: communication energy and computation energy. Communication energy forms the major contributor of the total energy consumption. Hence, in this work communication energy consumption analysis is performed. Communication energy is the sum of transmission energy and reception energy which is computed using Eqs. (15) and (17). The communication energy for the foreground measurements are computed in real time using powertrace tool in ContikiOS [36].

Transmission energy
Energy for transmitting the foreground measurements through TelosB nodes are calculated theoretically using Eq. (15)

$$E_{tx} = (t * I_t * V)/1024 \text{ J} \tag{15}$$

where t represents the time to transmit a 128 byte packet, which is approximately $1/250$ s, I_t represents the current, which is 19.5 mA, and V represents the voltage, which is 3 V. The current and voltage values are taken from the TelosB datasheet [37]. TelosB has a MSP 430f1611 processor with 10 K RAM, 48 K FLASH and 1 MB external flash [37]. The transmission energy per frame E_f is calculated using Eq. (16).

$$E_f = B_M * E_{tx} \tag{16}$$

where B_M represents the number of bits to be transmitted and $E_{tx} = 0.23 \, \mu\text{J}$ is calculated using (15) [5].

Reception energy
Energy for receiving per bit in TelosB mote is computed using Eq. (17)

$$E_{rx} = (t * I_r * V)/1024 \text{ J} \tag{17}$$

Table 4 Comparison of proposed approach with existing approach in terms of detection accuracy and energy

Approaches	Foreground measurements	Detection accuracy	Communication energy per frame (mJ)	
			Theoretical	Practical
CS based background subtraction	36,432	0.35	146	166.4
Mean measurement differencing with threshold strategy	11,077	0.83	44.4	50.6

where the values of t, I_r, and V are taken from the TelosB datasheet. The energy for receiving per bit is calculated as $0.27\,\mu J$. Table 4 shows the theoretical and practical computation of communication energy for proposed approach with threshold strategy and CS based background subtraction approach.

Table 4 shows PSO based EADS with the proposed object detection approach achieving 70 % reduction in communication energy compared to the existing CS based background subtraction approach. The mean measurement differencing approach yields 58 % higher detection accuracy and 89 % reduction in samples which in turn increases the network lifetime.

5.3 Network Lifetime

Network lifetime is analyzed for CS based background subtraction with PSO and the proposed mean measurement differencing with PSO methods by assuming a random network deployment. Communication energy alone is used for the analysis. The source node is chosen to be at the centre and the destination node is chosen randomly. The routing path is calculated using PSO with the parameters given in Table 2 and a suitable objective function is given in Eq. (18)

$$Path_{opt} = min(energy) \tag{18}$$

Communication Energy (ECOM)
The energy consumed for 'per bit' transmission and the energy consumed for 'per bit' reception are given in Eqs. (19) and (20), respectively.

$$E_{TX} = \varepsilon_e + \varepsilon_a d^\alpha \tag{19}$$
$$E_{RX} = \varepsilon_e \tag{20}$$

Fig. 16 Simulation scenario
for network lifetime

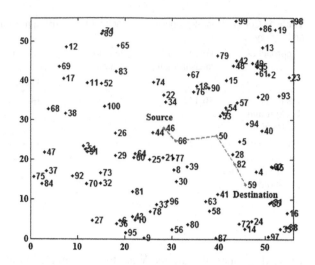

where $\varepsilon_e = 100 \times 10^{-12}$ is the energy dissipated per bit per m^2, $\varepsilon_a = 50 \times 10^{-9}$ is the energy consumed by the circuit per bit, d is the distance between a wireless transmitter and a receiver and $\alpha = 2$ is the path loss parameter [38].

The communication energy per bit is calculated as.

$$E_{COM} = E_{TX} + E_{RX} \tag{21}$$

The total energy consumption in multihop transmission is given below,

$$E_{TOT1} = E_{COM\_path} \times Bits_{tot} \tag{22}$$

The communication energy per bit for the entire path is multiplied by the total bits transmitted (Bits$_{tot}$) to get the total transmission energy.

For a sample scenario, nodes are randomly placed in a square area of 56 m × 56 m with an average node degree of 10. The number of nodes used is 100. The node closest to the centre of the field is chosen as the source node and the destination node is chosen randomly. The destination nodes are chosen such that there are four hops between the source and destination. The optimal path obtained using PSO is shown in Fig. 16.

The parameters of the simulated network are given in Table 5.

Each node in the network is provided with 10,800 J of energy initially as provided by the battery. Transmission of an image from the source to destination is considered as a session. The simulation ends when anyone of the nodes in the network is drained. The system lifetime is measured in terms of the number of sessions for the raw frame transmission, CS based background subtraction with PSO method and proposed mean measurement differencing with PSO method. Lifetime is also calculated by including the PVEH into account. The relay nodes are assumed to have an initial

Table 5 Simulated network parameters

Parameter	Value
Network size	56 m × 56 m
Number of nodes	100
Initial energy of nodes	
Without harvester	10,800 J
With harvester	11,993.536 J

Table 6 Network lifetime analysis

Method	Lifetime without harvester (sessions)	Lifetime with harvester (sessions)
Mean measurement differencing with threshold strategy	1,071,373	1,189,720
CS based background subtraction	344,274	382,303
Raw frame transmission	123,724	137,391

energy of 11,993.536 (battery + harvester energy). The lifetime is calculated in terms of sessions for all the methods and the results are listed in the Table 6.

Table 6 shows that the proposed method have 67.9 and 88.45% increased lifetime when compared to CS based background subtraction and raw frame transmission respectively. With solar energy harvester the proposed method achieves 10% increase in the lifetime for a single recharge cycle. Considering the number of recharge cycles throughout entire lifetime of the battery, the batteries would achieve an everlasting lifetime with solar energy harvester.

6 Conclusion and Scope for Future Work

An efficient anomaly detection system is proposed for WVSN which can detect the presence of an anomaly and alert the network operator. CS is used for addressing the energy and bandwidth constraints of the sensor nodes and PSO is used to optimize the minimum number of CS measurements which in turn reduces the memory, energy and bandwidth to a great extent while preserving detection accuracy. A mean based differencing approach has been proposed for EADS to obtain the differenced measurements from the background and current frames. A threshold is designed for extraction of the foreground measurements from the differenced measurements. These measurements are transmitted via TelosB nodes through the optimized routing path obtained from PSO. The performance of the EADS shows that the proposed differencing approach yields better detection accuracy compared to the existing CS based background subtraction approach. The proposed method has 67.9 and 88.45% increased lifetime when compared to CS based background subtraction and raw frame transmission respectively. To improve the network lifetime a solar harvester is designed which when used with TelosB nodes increases the network lifetime up to 10% for a single recharge cycle. Considering the number of recharge cycles throughout entire lifetime of the battery, the batteries would achieve an everlasting lifetime with solar energy harvester.

In future the PC at the transmitter side will be replaced by camera capable sensor mote with higher processing capability to implement the PSO based EADS system along with the optimized network analyzed in this work to transmit the information to the network operator. The latency involved in the transmission will also be analyzed.

References

1. Ye, Y., et al.: Wireless video surveillance: a survey. IEEE Access **1**, 646–660 (2013)
2. Baraniuk, R.: A lecture compressive sensing. IEEE Signal Process. Mag. **24**(4), 118–121 (2007)
3. Candes, E.J.: Compressive sampling. In: Proceedings of the International Congress of Mathematicians. European Mathematical Society, Madrid (2006)
4. Hemalatha, R., Radha, S., Sudharsan, S.: Energy-efficient image transmission in wireless multimedia sensor networks using block-based. Compress. Sens. Comput. Electr. Eng. **44**, 67–79 (2015)
5. Aasha Nandhini, S., et al.: Video compressed sensing framework for wireless multimedia sensor networks using a combination of multiple matrices. Comput. Electr. Eng. **44**, 51–66 (2015). doi:10.1016/j.compeleceng.2015.02.008
6. Wang, Q., Liu, Z.: A robust and efficient algorithm for distributed compressed sensing. Comput. Electr. Eng. **37**(6), 916–926 (2011)
7. Lei, J.: Generalized reconstruction algorithm for compressed sensing. Comput. Electr. Eng. **37**(4), 570–588 (2011)
8. Hassanien, A.-E., Abraham, A. (eds.): Computational Intelligence in Multimedia Processing: Recent Advances, vol. 96. Springer, Berlin (2008)
9. Kulkarni, R.V., Forster, A., Venayagamoorthy, G.K.: Computational intelligence in wireless sensor networks: a survey. IEEE Commun. Surv. Tutor. **13**(1), 68–96 (2011). (First Quarter)
10. Mendis, C., Guru, S.M., Halgamuge, S., Fernando, S.: Optimized sink node path using particle swarm optimization. In: Advanced Information Networking and Applications, 2006. AINA 2006. 20th International Conference on, vol. 2(5), pp. 18–20 (2006). doi:10.1109/AINA.2006.254
11. Ngatchou, P.N., Fox, W.L.J., El-Sharkawi, M.: Distributed sensor placement with sequential particle swarm optimization. In: IEEE Proceedings on Swarm Intelligence Symposium, SIS (2005)
12. Salas, V., Barrado, A., Lazaro, A.: Review of the maximum power point tracking algorithms for stand-alone photovoltaic systems. Solar Energy Mater. Solar Cells **90**, 1555–1578 (2006)
13. Sarvi, M., Ahmadi, S., Abdi, S.: A PSO-based maximum power point tracking for photovoltaic systems under environmental and partially shaded conditions. Prog. Photovolt. Res. Appl. **23**(2), 201–214 (2015)
14. Abdulkadir, M., Yatim, A.H.M., Yusuf, S.T.: An improved PSO-based MPPT control strategy for photovoltaic systems. Int. J. Photoenergy (2014), Article ID 818232
15. Joseph, P., Robert, S., David, C.: Telos: enabling ultra-low power wireless research. In: Proceedings of 4th International Symposium on information Processing in Sensor Networks, Los Angeles, CA, pp. 364–369 (2005)

16. Yi, Z., Liangzhong, F.: Moving object detection based on running average background and temporal difference. In: International Conference on Intelligent Systems and Knowledge Engineering (ISKE), pp. 270–272. IEEE (2010)
17. Cevher, V., Sankaranarayanan, A., Duarte, M.F., Reddy, D., Baraniuk, R.G., Chellappa, R.: Compressive sensing for background subtraction. In: Computer Vision—ECCV 2008, pp. 155–168. Springer, Berlin (2008)
18. Chen, S., Donoho, D.: Basis pursuit. In: Signals, Systems and Computers. 1994 Conference Record of the Twenty-Eighth Asilomar Conference on, vol. 1. IEEE (1994)
19. Tropp, J., Gilbert, A.: Signal recovery from partial information via orthogonal matching pursuit. IEEE Trans. Inf. Theory 53(12), 4655–4666 (2007)
20. Blumensath, T., Davies, M.E.: Iterative hard thresholding for compressed sensing. Appl. Comput. Harmonic Anal. 27(3), 265–274 (2009)
21. Needell, D., Tropp, J.A.: CoSaMP: iterative signal recovery from incomplete and inaccurate samples. Appl. Comput. Harmonic Anal. 26(3), 301–321 (2009)
22. Chartrand, R., Yin, W.: Iteratively reweighted algorithms for compressive sensing. In: Acoustics, Speech and Signal Processing, 2008. ICASSP 2008. IEEE International Conference on, pp. 3869–3872. IEEE (2008)
23. Haupt, R.L., Haupt, S.E.: Practical Genetic Algorithms. Wiley, New York (2004)
24. Jeyakumar, D.N., Jayabarathi, T., Raghunathan, T.: Particle swarm optimization for various types of economic dispatch problems. Int. J. Electr. Power Energy Syst. 28(1), 36–42 (2006)
25. Yang, S., Zhu, W., Chen, L.: Particle swarm learning algorithm based on adjustment of parameter and its applications assessment of agricultural projects. In: Computer and Computing Technologies in Agriculture II, vol. 2, pp. 1379–1388. Springer, Berlin (2008)
26. Zhang, J., Wang, B., Zhang, X., Huang, D.S., Zhang, X., Reyes García, C., Zhang, L.: Advanced Intelligent Computing Theories and Applications. Springer, Berlin (2010)
27. Mini, S., Udgata, S.K., Sabat, S.L.: Sensor deployment and scheduling for target coverage problem in wireless sensor networks. IEEE Sens. J. 14(3), 636–644 (2014)
28. Eberhart, R.C., Kennedy, J.: A new optimizer using particle swarm theory. In: Proceedings of the Sixth International Symposium on Micro Machine and Human Science, pp. 39–43 (1995)
29. Walker, G.: Evaluating MPPT converter topologies using a Matlab PV model. J. Electr. Electron. Eng. 21(1), 49–56 (2001)
30. Tremblay, O., Dessaint, L.A.: Experimental validation of a battery dynamic model for EV applications. World Electric Veh. J. 3, 1–10 (2009)
31. www.nrel.gov/
32. 'ContikiOS', http://www.contiki-os.org
33. http://www.wisdom.weizmann.ac.il/~vision/SpaceTimeActions.html
34. Cao, Y., Lei, Z., Huang, X., Zhang, Z., Zhong, T.: A vehicle detection algorithm based on compressive sensing and background subtraction. AASRI Proc. 1, 480–485 (2012)
35. Smitha, H., Palanisamy, V.. Detection of stationary foreground objects in region of interest from traffic video sequences. Int. J. Comput. Sci. Issues (IJCSI) 9(2) (2012)
36. Dunkels, A., Eriksson, J., Finne, N., Tsiftes, N.: Powertrace: Network-Level Power Profiling for Lowpower Wireless Networks, Technical Report T2011:05. SICS (2011)
37. http://www.memsic.com/userfiles/files/Datasheets/WSN/telosb_datasheet.pdf
38. Huaming, W., Alhussein, A.A.: Energy efficient distributed image compression in resource constrained multihop wireless networks. Comput. Commun. 28, 1658–1668 (2005)

Planning Robust Sensor Relocation Trajectories for a Mobile Robot with Evolutionary Multi-objective Optimization

Benjamin Desjardins, Rafael Falcon, Rami Abielmona and Emil Petriu

Abstract Wireless sensor networks provide a method for monitoring a region of interest. Incorporating a mobile robot within the sensor network allows various types of functionality to be added. One example of this is the replacement of risky and/or damaged sensors with other functional, passive ones. Using a specially designed *risk management framework* (RMF), we can proactively detect sensors that are at a high risk for failure and replace them before any network coverage is lost. The problem of optimizing the robot trajectory while picking up passive sensors and dropping them at the locations of the damaged sensors in the field has been studied as the *"Robot-Assisted Sensor Relocation"* (RASR) problem. One shortcoming of existing RASR methods is that the chosen robot trajectory is the one with the shortest length; however, no regards as to the durability of the passive sensors in the relocation chain are taken into consideration. We propose a more robust manner to come up with these trajectories by taking into account the current energy levels of the participating passive sensors as well as the ideal locations for their deployment. We resort to multi-objective optimization (MOO) to handle the tradeoffs among the different decision objectives that are part of this new formulation, named here as "Reliable Robot-Assisted Sensor Relocation". We outline the RRASR problem as well as the RMF used for detecting risky sensors in the wireless sensor network before the calculation of the sensor relocation trajectory takes place. We also evaluate the performance of six state-of-the-art evolutionary *multi-objective optimization* (EMOO) algorithms with sensor networks of varying sizes, inflicted damage levels, and passive sensor

R. Falcon (✉) · R. Abielmona
Research and Engineering, Larus Technologies Corporation, 170 Laurier Ave West - Suite 310, Ottawa, ON, K1P 5V5, Canada
e-mail: rfalcon@ieee.org, rafael.falcon@larus.com

R. Abielmona
e-mail: rabielmo@ieee.org, rami.abielmona@larus.com

B. Desjardins · E. Petriu
School of Electrical Engineering and Computer Science, University of Ottawa,
800 King Edward Ave, Ottawa, ON, K1N 6N5, Canada
e-mail: bdesj038@uottawa.ca

E. Petriu
e-mail: petriu@uottawa.ca

© Springer International Publishing AG 2017
A. Abraham et al. (eds.), *Computational Intelligence in Wireless Sensor Networks*,
Studies in Computational Intelligence SCI 676, DOI 10.1007/978-3-319-47715-2_8

densities. The empirical results confirm the feasibility of utilizing EMOO approaches to suggest multiple sensor relocation trajectories to the network manager.

Keywords Sensor relocation · Wireless sensor and robot networks · Multi-objective optimization · Wireless sensor networks · Genetic algorithms

1 Introduction

A wireless sensor is a device that is equipped with monitoring hardware, a battery, and some form of wireless communication [1]. A group of these wireless sensors is referred to as a *Wireless Sensor Network* (WSN) and is deployed in a region of interest (ROI) [1]. WSNs gather low-level information about a ROI for high-level applications in various domains, including medicine, defense, and agriculture [1]. As many of these domains are information-critical, the sensors must be deployed in such a way to provide complete sensing coverage over the ROI with no internal sensing holes.

WSNs can be functionally expanded with the addition of a mobile robot. This robot, or group of robots, perform some sensing or actuation task that is related to the WSN. This is referred to as a *Wireless Sensor and Robot Network* (WSRN) [2]. The robots add a flexibility not present in a WSN as they are generally mobile and have fewer or no resource restrictions and are designed to maintain, assist, or optimize the sensor network. Falcon explains in [2] that WSRNs can be divided into two categories: *robot-dependent WSNs*, and *robot-assisted WSNs*. In a robot-dependent WSN, the robot(s) are integral to the functionality of the network. However, robot-assisted WSNs do not require the functionality of robot(s), instead only relying on them to improve the performance of the network.

Robot-Assisted Sensor Relocation (RASR) is a specific WSRN-related problem wherein a robot is responsible for maintaining the coverage of the network. RASR is also referred to as *"carrier-based coverage repair"* [3] in the literature. This problem is defined by having a common base station where one or more robots are on standby. These robots are responsible for relocating and/or replacing sensors in order to maximize network performance. RASR assumes that there are additional sensors in the ROI not currently being used by the network to achieve maximum coverage. These extra sensors do not actively collect information and will be referred to as *passive sensors*. The robot(s) use these sensors to replace damaged sensors in the network, viz. by leaving the base station, picking up some passive sensors and dropping them at the locations of the damaged sensors to reconfigure the network when there has been enough degradation of the coverage area.

This chapter builds upon the early work done by Desjardins et al. [4] in defining *Reliable Robot-Assisted Sensor Relocation* (RRASR). The original work defines RRASR as a specific version of RASR wherein there is a single mobile robot that has a limited carrying capacity and our objective is to find an optimal sensor relocation trajectory. RRASR differs from RASR in that there are additional objectives of importance beyond trajectory length. These additional objectives are representative of the fact that when only considering the distance that the mobile robot travels, we ignore the quality of the sensors that are being placed. In the worst case scenario, a passive sensor with very little remaining battery life could be used to fill a sensing hole. This is a poor solution as it will require the robot to redeploy in a short amount of time to fix the same sensing hole. Due to these additional objectives, RRASR was formulated as a *multi-objective optimization* (MOO) problem and as such we use *evolutionary multi-objective optimization* (EMOO) algorithms to solve it. The decision to solve the problem as an explicit multi-objective problem stems from the desire to provide a decision maker with a variety of good solutions so that they may choose one based on the situation at hand.

In the original RRASR problem [4], robots are deployed when there are sensing holes already in the network. In this chapter, we present a method for identifying sensors that are likely to fail, and replacing them before a sensing hole occurs. This is essential for operations in critical systems where complete sensing coverage of an ROI is always required. We rely on the *Risk Management Framework* (RMF), described in [5] to support the risk-aware analysis of which sensors become damage at a certain point in time. Our tailored RMF formulation allows proactive replacement of sensors that are deemed at risk of failing or being faulty in order to try and minimize any lack of coverage in the ROI.

We expand on the original work in [4] with the following contributions: (1) we review the literature on three fronts: sensor relocation by mobile robots, the optimization problem that stems from it, and the use of risk-driven schemes; (2) we propose a proactive replacement methodology using an RMF-based approach that monitors several key factors to ensure proper sensor functioning; (3) we modify one of the existing RRASR objectives proposed in [4] to more accurately reflect the importance of network connectivity when selecting what passive sensor to deploy at a certain spot; (4) we solve this problem using a set of state-of-the-art algorithms, including the newer AGE-I [6] and AGE-II [7] algorithms; (5) we use parametric tuning to ensure each algorithm is providing a set of good solutions; (6) we discuss on our statistically validated empirical results, giving insight into the EMOO algorithms' performance, in experiments dealing with various sizes of sensor networks, inflicted damage levels, and distribution density of passive sensors.

The remainder of the chapter is structured as follows: Sect. 2 briefly reviews relevant works. Section 3 details our proposed RMF formulation for determining the damaged sensors, Sect. 4 formalizes the RRASR problem and Sect. 5 elaborates on the algorithmic components for the EMOO techniques under consideration. Section 6 contains the empirical evaluation of the proposed methodology. Finally, Sect. 7 concludes the chapter.

2 Related Work

This section provides a brief overview of related studies regarding robot-assisted sensor relocation and its underlying optimization problem as well as the risk management framework.

2.1 Robot-Assisted Sensor Relocation

RASR has been studied in various forms using different methodologies, as summarized in Table 1. In addition to what method is used to solve the problem, there are two other categories that define an approach to RASR: the architecture, centralized or localized, and the number of robots being one or many. The formalizations provided in [3, 8, 9, 13] use a centralized architecture, wherein a central server or base is responsible for determining the deployment and trajectory of the robot(s). In the centralized scenario, the robot(s) are stored in the base station until they are needed. Multiple-robot solutions are described in [9, 11, 14] and use different methods for determining trajectories of the robots. All of the literature examined provided solutions where the robot has a limited carrying capacity. Uniquely, the solution provided by Li et al. [14] uses a method where the sensors themselves play a role in helping the robot to identify sensing holes in the network. Fletcher et al. [12] use a strictly programmatic approach to replacement whereby one or many robots will wander a WSN and replace sensors as necessary. They provide two variants of their method: grid-based and non-grid-based.

Within the RASR space, the method closest to the one we propose is found in [13]. Coincidentally, it also differs the most from the others as it proposes a method where only the batteries of the sensors are replaced, rather than the sensors themselves. It uses a genetic algorithm (GA) to solve a two objective MOO problem wherein one objective is favoured over the other. One of the objectives represents the portion of damaged sensors that are in the current solution. This acts as a driving factor for all damaged sensors to be included in the solutions, but does not add depth to the problem.

Our method provides more depth to the problem by more realistically modeling it. We do this by incorporating an MOO approach that stems from the consideration given to other facets of the problem. We extend the problem solved in [3] by including additional decision objectives for determining the optimal sensor relocation trajectories. We set ourselves apart from traditional RASR solutions by incorporating additional decision objectives that provide a more realistic view of RASR.

Table 1 Related RASR literature

Ref.	Architecture	Number of robots	Methodology	Assumptions
[3]	Centralized	One	Ant Colony System	Robot has no battery constraints Robot can only carry a limited number of sensors
[8]	Centralized	One	Ant Colony System + Constrained Neighbourhood Mutation (ACS+CNM)	Robot has no battery constraints Robot can only carry a limited number of sensors
[9]	Centralized	Many	Firefly Algorithm and Harmony Search	Robot has no battery constraints. Robot can only carry a limited number of sensors
[10]	Localized	One	Proactive and reactive replacement algorithms	Robot may only carry one sensor Robot has battery life Robot wanders when not recharging
[11]	Localized	Many	Localized Ant-based Sensor Relocation Algorithm with Greedy Walk (LASR-G)	Robot may only carry one sensor Maximizes coverage in any situation be it node failure or poor deployment
[12]	Localized	One or Many	Randomized Robot-assisted Relocation of Static Sensors (R3S2) and grid variant (G-R3S2)	Robot has no battery constraints Robot wanders when not actively replacing nodes
[13]	Centralized	One	Dual-objective genetic algorithm	No relocation of nodes, only replacing batteries Batteries deplete over the course of replacement
[14]	Localized	One or Many	Market-based Sensor Relocation (MSR)	Robot may only carry one sensor

Table 2 Related optimization literature

Ref.	Optimization problem	Optimization algorithm	Objective	Assumptions
[15]	Selective Pickup and Delivery Problem (SPDP)	Genetic Algorithm (GA)	Minimize route length	No maximum vehicle load
[16]	Multi-vehicle SPDP	GA with path relinking	Minimize sum of route lengths	All vehicles are used in each instance
[17]-1	SPDP	GA with adaptive mutation	Minimize route length	No maximum vehicle load
[17]-2	SPDP	GA with local search and adaptive mutation	Minimize route length	No maximum vehicle load
[18]	Single Vehicle Routing Problem with Deliveries and Selective Pickups (SVRPDSP)	GA with variable neighbourhood search	Minimize route length	Multiple visits to delivery location is valid. Allows pickup and delivery from same location
[19]	SVRPDSP	Evolutionary algorithm with variable neighbourhood descent	Minimize route length	Multiple visits to delivery location is valid. Allows pickup and delivery from same location
[20]	SPDP	GA with local search	Minimize route length	No maximum vehicle load
[21]	Multiple Vehicle Routing Problem with Delivery and Selective Pickups (MVRPDSP)	Hybrid metaheuristic	Minimize sum of route lengths	Allows pickup and delivery from same location

2.2 Related Optimization Problems

RASR can be generalized into versions of classical optimization problems: the Travelling Salesman Problem (TSP) and the Vehicle Routing Problem (VRP). The versions of these problems that most closely model RASR are *selective pickup and delivery problem* (SPDP) [15] and the *Single Vehicle Routing Problem with Deliveries and Selective Pickups* (SVRPDSP) [18]. The defining difference between TSP/ VRP and SPDP/ SVRPDSP is that all delivery requests must be satisfied, but not all pickup locations need to be visited. Table 2 gives an overview of some of the approaches to these problems. As a note, since TSP and VRP are known to be NP-hard, we can infer that RASR is NP-hard as well.

These problems have been approached in different ways, but of interest to us are approaches using evolutionary and genetic algorithms. Liao and Ting use a standard GA approach in [15]. An important feature of this work is that they incorporate a repair function in order to guarantee the feasibility of their solutions. This approach is expanded upon in [20] where they use a memetic algorithm in order to improve their results. In [17], Liao and Ting employ both a standard GA approach and a memetic approach with adaptive mutation to enhance solution quality. In [18], Bruck et al. use a GA with variable neighbourhood search. This is further developed in [21] where a multi-vehicle version of the problem is examined. The multi-vehicle problem is also examined by Huang and Ting in [16]. The approach in [18] is broadened on by Bruck and dos Santos in [19] where they explore a different gene representation as well as incorporate data mining techniques in order to improve their mutation and crossover operators.

Our approach diverges from the ones mentioned above due to the addition of multiple decision objectives. As the methods relating to this problem in the literature only address a single objective, we unveil a new avenue to the problem. As we consider using EMOO algorithms to solve our problem, the genetic representations found in [15–20] can still be deemed relevant as it is possible to use them in an EMOO algorithm.

2.3   Risk-Driven Detection of Damaged Sensors

The original RMF was proposed by Falcon et al. [5]; the authors outlined a multi-modular architecture comprising: (a) a risk feature extraction module that generates a parallel risk stream from the incoming raw data features; (b) a risk visualization module, allowing the user to monitor the system's risk landscape in real time and (c) a risk assessment module that evaluates local and global risk dimensions of any system unit at any point in time. In [22], a response selection module was added in order to automatically determine a small set of the most promising responses to a risky event that are to be presented to an operator for further analysis. McCausland et al. give in [23, 24] two implementations of an RMF for maintaining perimeter coverage of an ROI. This is similar to our proposed RMF as we too are trying to preserve maximum coverage across our ROI. Falcon et al. augment the original risk feature extraction module in [25] by integrating hard-soft information fusion in order to extract risk features from a variety of sources such as radar, Automatic Identification System (AIS), intelligence reports, and historical data. More recently, the generation of an object's intent in a risk-aware fashion using anomaly detectors has been the subject of discussion in [26].

3 RMF for Sensor Fault Detection

In the original work [4] the robot was only deployed when the sensing holes in the ROI had formed due to sensor failure caused by battery depletion or otherwise. Here, we outline outline an implementation of the Risk Management Framework introduced in [5] that aims at detecting damaged sensors early on so they could be appropriately replaced with passive sensors. to maintain a maximal amount of sensor coverage over the ROI. We incorporate elements of the RMF proposed in [22] including *risk feature extraction* and the *risk assessment module*.

Risk feature extraction works by identifying the underlying data features of our system and providing a transformation into risk features used by the *risk assessment module*. This is done by creating a set of linguistic terms for each linguistic variable. Each linguistic term is modeled as a fuzzy set with a corresponding membership function. We use a Mamdani fuzzy inference system [27] for our implementation of the risk assessment module.

3.1 Data Features

We use the following four data features as the basis for our RMF:

Battery Level A value reported from the sensor, in the range [0, 100] representing the percentage of battery power remaining.

Data Fault Detection A nominal value representing the quality of data being transmitted by a sensor. It is determined by the method proposed by Chen et al. [28]. The nominal value has four possibilities: *possibly normal* (LG), *possibly faulty* (LT), *normal* (GD), or *faulty* (FT). GD and FT indicate sensors that are positively normal or faulty respectively whereas LT and LG indicate the lack of required information to place the sensor into either the GD or FT categories.

Transmission Reliability Used to measure if the frequency at which data is periodically sent by a sensor is acceptable. This is represented by the following function:

$$f(r) = r/T_{min} \tag{1}$$

where r is the current transmission rate based on a sliding time window, in packets per second, and T_{min} is the minimum acceptable transmission rate. If $f(r) >= T_{exp}$, where $T_{exp} = \rho T_{min}$, the expected transmission rate, and ρ is a predetermined constant based on the hardware used, then the transmission rate is deemed to be normal. If $1 < f(r) < (\rho + 1)$ then the transmission rate is acceptable.

Physical Reliability Represents the likelihood of a failure occurring due to the age of the sensor or its components. We represent this with the exponential failure distribution function:

$$F(t) = 1 - e^{-\lambda t} \tag{2}$$

where t is the usage time of the sensor and λ is a constant that is determined by the make and model of the sensor.

3.2 Risk Features

Each of the data features described in Sect. 3.1 corresponds to a risk feature. This gives us the following risk features:

1. Battery Risk (R_1)
2. Data Fault Risk (R_2)
3. Transmission Reliability Risk (R_3)
4. Physical Reliability Risk (R_4)

Each of the risk features is treated as a linguistic variable with three linguistic terms: *LOW*, *MED*, and *HIGH* to indicate the level of risk that the data feature entails. Using the functions described for each data feature, we assign a risk value for each of the risk features by using trapezoidal membership functions. These functions are detailed in Table 3 for R_1, R_3, and R_4. These functions are designed as a proof-of-concept and serve to provide a basis for experimentation. R_2 is a unique case as it

Table 3 Trapezoidal membership functions

	Battery risk			
	A	B	C	D
High	$-\infty$	0	15	30
Med	20	30	60	70
Low	60	75	100	∞
	Transmission reliability risk			
High	$-\infty$	0	1	$1 + 0.25(\rho - 1)$
Med	$1 + 0.1(\rho - 1)$	$1 + 0.6(\rho - 1)$	$1 + 0.75(\rho - 1)$	$1 + 0.85(\rho - 1)$
Low	$1 + 0.75(\rho - 1)$	$1 + \rho$	∞	∞
	Physical reliability risk			
High	0.6	0.85	1	∞
Med	0.3	0.5	0.6	0.8
Low	$-\infty$	0	0.33	0.5

is defined by four nominal values; as such, we assign the value of the risk features according to the following:

$$\text{Possibly Normal (LG)} \mapsto LOW$$
$$\text{Normal (GD)} \qquad \mapsto LOW$$
$$\text{Possibly Faulty (LT)} \; \mapsto MED$$
$$\text{Faulty (FT)} \qquad \mapsto HIGH$$

Using the membership values for the three numerical risk features and the nominal value for the only nominal risk feature, we are able to determine an overall risk value for each active sensor in the network. This process is outlined in the following section.

3.3 Risk Assessment Module

The assessment module of the RMF aggregates the risk features into an overall risk value for each sensor. Our RMF uses the following rules for aggregation:

- IF R_1 is *HIGH* or R_2 is *HIGH* or R_3 is *HIGH* or R_4 is *HIGH* then *SensorRisk* is *HIGH*
- IF R_1 is *MED* and R_2 is *MED* and R_3 is *MED* then *SensorRisk* is *HIGH*
- IF R_1 is *MED* or R_2 is *MED* or R_3 is *MED* or R_4 is *MED* then *SensorRisk* is *MED*
- IF R_1 is *LOW* and R_2 is *LOW* and R_3 is *LOW* and R_4 is *LOW* then *SensorRisk* is *LOW*

If a sensor belongs to *HIGH* with membership greater than 0.6 a replacement operation is triggered. Any sensor with membership to $MED \geq 0.6$ or $HIGH \geq 0.6$ is flagged for replacement by the mobile robot. We replace sensors with high membership to *MED* in order to reduce the number of relocation tours required by the robot to maintain network coverage. Note that some of the risk features do not increase monotonically, so threshold values for flagging must be carefully tuned to avoid replacing sensors that do not necessarily require replacement. The rules given here have been designed in order to reduce the amount of coverage loss to a minimum (Fig. 1).

4 RRASR: A Multi-objective RASR Formulation

RRASR can be expressed as a combinatorial optimization problem by representing our sensor network as a complete undirected graph $G = (V, E)$ with vertex set $V = \{v_0, \ldots, v_n\}$ and edge set $E = \{e_{ij} = (v_i, v_j) | v_i, v_j \in V, v_i \neq v_j\}$ in which each edge e_{ij} has a cost $d_{ij} > 0$ representing the Euclidean distance between v_i and v_j.

Each vertex represents either a passive sensor or a sensing hole and has an associated unitary demand q_i (1 for passive sensors and -1 for sensing holes). The base station is denoted as v_0 with $q_0 = 0$. Therefore, we can say that any other vertex

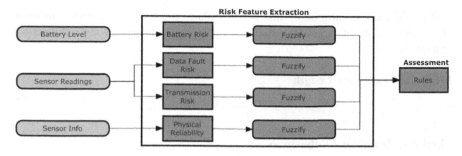

Fig. 1 An outline of the steps of the RMF for RRASR

either belongs to the set of passive sensors $S = \{v_i | v_i \in V, q_i = 1\}$ or the set of sensing holes $H = \{v_i | v_i \in V, q_i = -1\}$. This implies that $V = \{S \cup H \cup v_0\}$.

Each sensor has an associated battery level denoted by:

$$p_i = \begin{cases} 0 & \text{if } v_i \in H \\ -1 & \text{if } i = 0 \\ \sim U(0, 100) & \text{if } v_i \in S \end{cases} \qquad (3)$$

Note that $\sim U(0, 100)$ refers to a random number uniformly distributed in the specified interval of 1 and 100.

There is a unique commodity (sensors) to be transported by the robot from one place to another. The robot can carry at most Q_{max} sensors and leaves the base station with an initial cargo Q_0, $0 \leq Q_0 \leq Q_{max}$. The robot always drops the most recently picked up sensor when it reaches a sensing hole due to simulated design constraints on the robotic arm and mobile platform.

We also consider the set of all active nodes $A = \{a_0, \ldots, a_m, v_0\}$ that form the WSRN deployed in the ROI, although these do not become part of the network graph G for optimization purposes. This is an essential consideration to our problem as it allows us to determine the importance of a node in the WSRN with regards to maintaining network connectivity. v_0 is specially included in A to account for those sensing holes only known to the base station.

All passive and damaged sensors $v_i \in V$, $1 < i < n$ have a communication radius of R. We use the edge set $E' = \{e'_{ij} = (a_i \in A, v_j \in H)\}$ to find the active node degree g_j of each sensing hole $v_j \in H$ defined as the number of active sensors within communication distance of the sensing hole. This is formally defined as: $g_j = |\{a_i \in A : d'_{ij} \leq R\}|$ and d'_{ij} stands for the Euclidean distance between a_i and v_j where $a_i \in A$, $v_j \in H$.

We want to find a feasible sensor relocation trajectory, $\varphi = \left(v_{\gamma_1}, v_{\gamma_2}, \ldots, v_{\gamma_{|\varphi|}}\right)$ where $\gamma : [1; |V| - 1] \rightarrow \{\gamma_i\}$, $v \notin V\{v_0\}$ is an injective function, that starts and ends at the base station. As stated by Falcon et al. [3], a sensor relocation trajectory is said to be feasible if it has no repeated nodes (other than the base station as its first and last element), repairs all sensor holes (i.e., drops a passive sensor at the location of each sensing hole), and never violates the robot's capacity constraint, Q_{max}. The quality of φ will be evaluated according to the following optimization objectives:

Minimize Trajectory Length

$$\sum_{e_{ij} \in \varphi} d_{ij} \tag{4}$$

Maximize Trajectory Robustness

$$\sum_{v_i \in \varphi \cap S, v_j \in \varphi \cap H} p_i g_j \delta_{ij} \tag{5}$$

Maximize Trajectory Lifetime

$$\min\{p_i\}, v_i \in \varphi \cap S \tag{6}$$

The trajectory robustness is representative of how likely a sensor hole would be to cause a disconnect in the WSN. It is expressed as a value based on the battery power of a given sensor and its score, g_j. This score is determined using the concept of *vertex separation* and the following equation:

$$g_j = 1 + \sum_{a \in A} \sum_{v_j \neq a, v_j \neq v_i, v_j \in C_{abi}} 1/|C_{abi}|, v_j \in A \tag{7}$$

where C_{abi} is the ith ab-separator [29], $a \in A$, and $b = v_0$ the depot.
The function δ_{ij} is evaluated as follows:

$$\delta_{ij} = \begin{cases} 1 & \text{if } v_i \in \varphi \cap S \text{ will be dropped off at the location of } v_j \in \varphi \cap H \\ 0 & \text{otherwise} \end{cases} \tag{8}$$

The trajectory lifetime shows the worst passive sensor power level in a sensor relocation trajectory. We want the selected passive sensors to have enough battery power so they can sustain operational demands without being depleted in a short period of time, hence becoming damaged nodes themselves.

5 EMOO Algorithms for RRASR

This section is devoted to outlining the different components of the multiobjective optimizers under consideration to tackle the RRASR problem.

5.1 Solution Encoding

We use a modified version of a permutation-based solution representation seen in [15]. For a problem instance with a set of n nodes we will create a permutation of size $n-1$; we do not include the base station, v_0, in our permutation as it is assumed to be the beginning and end of every tour. Each solution will have the form $(v_1, \ldots, v_n)$ to represent the order in which the nodes are visited. As not all nodes are required to be visited, we will represent the unvisited nodes by marking them as negative integers. For example, the permutation $(6, 2, -4, 1, 3, -5)$ indicates that the robot will visit nodes 6, 2, 1, and 3, in order while ignoring the others. Since RRASR must have solutions computed quickly in order to act upon them in the WSN, we adopted this representation as it ensures higher computational efficiency at the expense of requiring greater memory capacity.

5.2 Objective Functions

As we are using an MOO approach to the problem, we use (4–6) as the fitness functions. As a result each solution is evaluated in terms of three different (and conflicting) objectives.

5.3 Population Initialization

The population is initialized by creating three heuristically determined individuals and randomly initializing the remainder of the population. The heuristically determined individuals are based on the three objective functions. For trajectory length (4) we use a nearest neighbour heuristic in which the shortest feasible edge is always taken. For trajectory robustness (5) we select the pickups and deliveries such that the best sensors are always delivered to the sensing hole with the highest score. For trajectory lifetime (6) we pickup the sensors with the highest battery levels, regardless of location. The random initialization of the remainder of the population is done by creating a permutation with no duplicate elements, with the sign (positive or negative) of each element being decided with equal probability.

5.4 Evolutionary Operators

Selection Operator We use a *binary tournament* as our selection operator, as it is the standard in the library used for implementation, the MOEA Framework.[1] This

---

[1] http://moeaframework.org/.

works by randomly selecting two individuals from the population and then selecting the best of the two. As this is an MOO problem, we use the non-domination rank [30] of a given solution rather than its fitness function values. In the case that the selected solutions share the same rank either solution is chosen with equal probability.

Crossover Operator Partially-mapped crossover (PMX) [31], as the default in our implementation framework, is used as the crossover operator. This type of crossover builds offspring solutions by selecting a subsequence from one of the parents and inserting it into the other parent, preserving the original order of as many points as possible, as demonstrated in Fig. 2.

Mutation Operator We include two mutation operators in our implementation: *insert* and *swap*. Each offspring has a chance to undergo a mutation. The *insert* operator works by selecting an element of the permutation and inserting it at another random location in the permutation and is illustrated in Fig. 3. The *swap* operator works by randomly selecting two elements of the permutation and switching their values and is illustrated in Fig. 4.

These mutations would have no or little effect if they were to only affect the negative elements of the permutation. As such, any element that is affected by a

Fig. 2 Example of partially-mapped crossover

Fig. 3 Example of the *insert* mutation operator

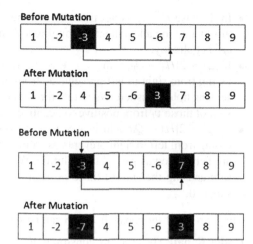

mutation also has its sign flipped. This causes the mutations to have a guaranteed effect on both the genotype and phenotype of the solution.

5.5 Infeasibility Handling

Solutions will be repaired if, upon evaluation, they violate any constraints. This is done so that the solution satisfies all of the problem constraints. Our proposed repair methodology can be broken down into two parts: *content repair* and *order repair*, which are completed in that order.

Content Repair In the case that the solution does not contain the appropriate number of passive sensors or does not contain all of the damaged sensors we use the following procedure:

Fig. 5 In the first example there are not enough passive sensors in the relocation trajectory. We add passive sensors from *right* to *left* until we have enough. In the second example there are too many passive sensors in the relocation trajectory. We remove passive sensors from *left* to *right*

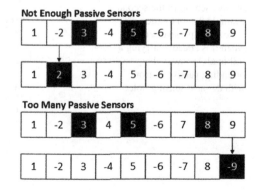

- In the case $H \subseteq \varphi$ is not true, we add the remaining elements of H to φ. We do this by making all elements belonging to H in the encoded solution positive if they are not already.
- If $|\varphi| > 2|H| - Q_0$, remove $v_i \in S \cap \varphi$ from right to left until $|\varphi| = 2|H| - Q_0$. Going from right to left prevents us from collecting passive sensors after the last sensing hole is visited. We iterate through the permutation from n to 1 and flip the sign of those v_i from positive to negative until the condition is satisfied.
- If $|\varphi| < 2|H| - Q_0$, add $v_i \in S$, $v_i \notin \varphi$ from left to right until $|\varphi| = 2|H| - Q_0$. Going from left to right prevents us from removing passive sensor collections that are being used as part of the solution. We iterate through the permutation from 1 to n and flip the sign of those v_i from negative to positive until the condition is satisfied.

This portion of the repair function can be seen in Fig. 5.

Order Repair The ordering constraint dictates that: at all points in φ, $0 \leq Q \leq Q_{max}$ must hold. We must consider the two infeasible cases for repair:

- If $Q > Q_{max}$ at some point in φ, the offending $v_i \in \varphi \cap S$ is moved after the nearest right $v_i \in \varphi \cap H$ in φ. To do this we insert the element in which the issue appears after the rightmost element of H that appears in the permutation.
- If $Q < 0$ at some point in φ, the offending $v_i \in \varphi \cap H$ is moved after the nearest right $v_i \in \varphi \cap S$ in φ. To do this we insert the element in which the issue appears after the rightmost element of S that appears in the permutation.

This portion of the repair function is illustrated in Fig. 6.

Fig. 6 Examples of sensor relocation trajectory ordering repair for $Q_{max} = 2$. In the first example the robot attempts to place a sensor when it is not carrying one. We swap the offending sensing hole with the next passive sensor. In the second example the robot attempts to collect a passive sensor when it is already carrying a full load. We swap the offending passive sensor with the next sensing hole

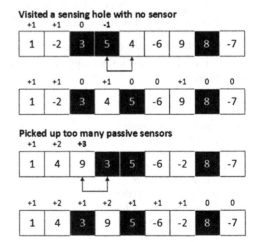

5.6 Stop Criteria

The algorithms will be terminated after they have undergone a maximum number of objective function evaluations (NFE). In a real-world deployment, it would be more appropriate to use a time-based stop criteria as RRASR applications will generally be time-constrained. We use objective function evaluations as our stop criteria as it gives us a larger view of algorithm performance independent of time, which is used as an evaluation metric.

6 Experimental Results

This section elaborates on the empirical methodology and evaluation for our proposed MOO methodology for the RRASR problem.

6.1 Experimental Setup

Synthetic Scenario Generation The scenarios used for experimentation have been generated under the control of various problem parameters such as: the total number of sensors N, ratio of sensing holes to passive sensors R_{HS}, and the distribution density of the passive sensors. In order to simulate an appropriate RRASR scenario, we generate a WSN by first mapping a grid of active sensors to a ROI. We then randomly select a subset of the active sensors to act as sensing holes. Finally, we randomly distribute the passive sensors using the *MAX_DPA* algorithm adapted from [32] with the distribution density parameter. The resulting collection of passive sensors, active sensors, and sensing holes provides the scenario used for experimentation.

We generated 110 scenarios in total: 50 scenarios for varying N from 10 to 1000 in increasing steps, 10 using varying distribution density values in *MAX_DPA* and 50 for differing R_{HS} values from 1 to 50 %.

Benchmark EMOO Algorithms We examine the performance of six different EMOO algorithms on our problem. We include NSGA-II [30], which is based upon Pareto dominance and a crowding distance operator, and its successor NSGA-III [33] which follows the same framework as NSGA-II, but also uses reference planes to drive towards good solutions. SPEA2 [34] uses a strength function that is based on the number of dominated solutions of the non-dominated set. PESA-II [35] uses a region-based approach whereby the solution space is broken up into regions and solutions are chosen in order to keep the best and maintain a healthy spread across regions. The newer AGE-I [6] and AGE-II [7] algorithms are included in our benchmark group and use a method that aims to have the population model approximate the solution space called *approximation-guided evolution*. AGE-II sets itself apart from

AGE-I by incorporating ε-dominance [36] and is stated to increase performance in regards to both solution quality and runtime.

Performance Metrics The performance metrics used in this study are outlined in Table 4. These are well-known MOO indicators that judge the goodness of an obtained Pareto approximation set from different angles. The reported metric values are the averages, over each synthetic RRASR scenario, of the means of 30 independent runs of an algorithm upon a scenario.

Algorithm and Parameter Configurations In order to provide a fair comparison of the algorithms we are testing, we must use the most appropriate parameters for each algorithm. To do this we conduct parametric tuning on each algorithm to determine the best parameter values, within a set range, for each algorithm. The parameters and their ranges are listed in Table 5. We generated 240 parameters vectors using the Saltelli method [39] and each algorithm was run 15 times for each vector, with

Table 4 Algorithm performance metrics

Name	References	Goal	Description
Hypervolume	[37]	Max	Represents the volume of the objective space dominated by solutions in the approximation set
Max Pareto front error	[37]	Min	Represents the maximum distance from solutions in an approximation set to the nearest solution in the reference set
Spacing	[37]	Max	Represents the spread of the Pareto approximation set
Generational distance	[37]	Min	Represents average distance from solutions in the approximation set to the nearest solution in the reference set.
Inverted generational distance	[38]	Min	Represents the average distance from solutions in the reference set to the nearest solution in the approximation set
Elapsed time	N/A	Min	The runtime (in seconds) of the algorithm

Table 5 Algorithm parameter ranges

Parameter	Min. value	Max. value
Number of function evaluations	50,000	200,000
Population size	50	500
Swap mutation rate	0.0	0.5
Insertion mutation rate	0.0	0.5
Crossover rate	0.5	1.0

Table 6 Tuned algorithm parameter values

Algorithm	NFE	Pop. size	Swap rate	Insert rate	Crossover rate
AGE-I	143,880	55	0.3716	0.2349	0.8628
AGE-II	172,040	110	0.4966	0.1099	0.9878
NSGA-II	195,545	185	0.0747	0.1255	0.9097
NSGA-III	181,390	110	0.3687	0.1353	0.9878
PESA-II	195,365	205	0.3843	0.4634	0.9097
SPEA2	195,545	185	0.3843	0.4634	0.7651

the results being averaged across all executions. We conducted this test over 14 of our synthetic data files that are representative of the set of problem instances under study.

To determine the best vector out of the available selection for each algorithm we collected the metrics listed in Table 4 and gave each vector a ranking from one (best) to 240 (worst) for each metric, then took the average of those ranks. The vector with the highest average rank across all metrics was deemed the best set of parameters for a given instance file. We then took the highest average vector from across all of the instance files. We did this by taking the average rank of each vector across all 14 data files. These parameter sets are used for the experiments detailed in Sects. 6.2–6.4 and are reported in Table 6.

Statistical Validation To confirm whether or not the differences in observed performance are statistically significant, we rely on nonparametric tests as suggested by [40]. We use 5 % significance for all tests.

Due to the novelty of our problem, we do not have a control algorithm. Therefore, we must first establish or reject statistically significant differences within the whole group of algorithms using the Friedman $N \times N$ procedure. In the case that a comparison yields a rejection of the null hypothesis (i.e. there exists a statistically significant difference), then a set of comparisons is done using the following post-hoc procedures: Bergmann et al. [40]. We first rely on the Bergmann procedure given its robustness and reliable adjustment of p-value. If the Bergmann procedure does not reject a hypothesis we will defer to the other post-hoc methods.

The algorithm performance data used for the statistical validation procedure was collected at 25, 50, 75, and 100 % of the number function evaluations across each of the scenarios.

6.2   Experiment 1: Scalability Analysis

This experiment examines the effect the size of the WSN has on the performance of the algorithms with regards to RRASR solutions. For this experiment we set the number of sensing holes to be 15 % of the total nodes.

Figures 7, 8 and 9 show the runtime (in seconds), hypervolume, and spacing metrics respectively against the size of the sensor network. Looking at the runtime we can see that as the size of the WSN and the solution space gets bigger the runtime increases as expected. It is important to note the difference in runtime for SPEA2, and to a lesser extent AGE-II. Unlike the other algorithms, SPEA2 and AGE-II have an uncharacteristically large runtime when the solution space is small. Looking at Fig. 8, we can see that this increase in runtime does not have a negative impact on solution quality. For SPEA2 this is likely the result of many duplicate solutions appearing in the archive it uses for the selection of parents, causing many unnecessary comparisons. AGE-II uses an *approximative archive* that contains only non-dominated solutions. As there are a finite number of non-dominated solutions, it is likely that they will quickly find their way into the archive causing many domination comparisons to be made to ensure that the archive only contains appropriate solutions.

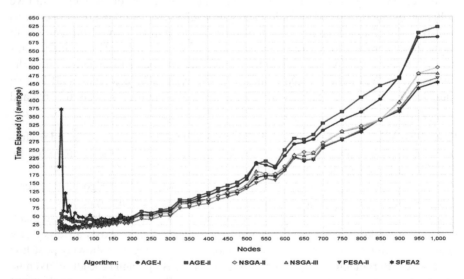

Fig. 7 Final execution results for time elapsed, in seconds, across all generated scenarios for the scalability analysis experiment

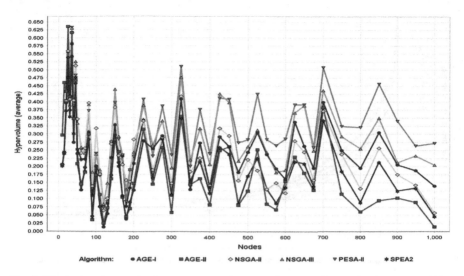

Fig. 8 Final execution results for hypervolume, across all generated scenarios for the scalability analysis experiment

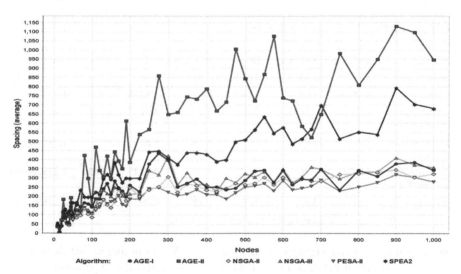

Fig. 9 Final execution results for spacing, across all generated scenarios for the scalability analysis experiment

In regards to overall algorithm performance we can see in Fig. 8 that PESA-II and NSGA-III consistently give the best solutions, in terms of hypervolume. The only downside to these solutions is that they may not have much variety, as we can see in Fig. 9 that these algorithms underperform the others in terms of spacing. While AGE-II has the most diverse set of solutions, they are also markedly the worst. This may be due to an increased focus on exploration of the solution space preventing

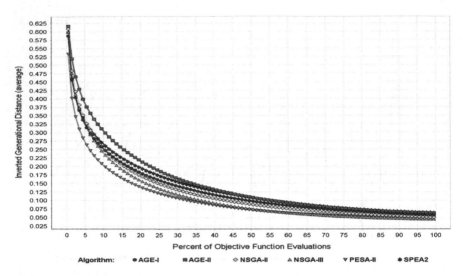

Fig. 10 Average results, across all generated scenarios, for the inverted generational distance for the scalability analysis experiment

Table 7 Experiment 1: Friedman $N \times N$ test results

	Algorithm rank						
Metric	NSGA-II	NSGA-III	SPEA2	PESA-II	AGE-I	AGE-II	p-value
Hypervolume	3.3175	**2.0125**	4.615	2.015	4.19	4.85	2.0159E−10
Spacing	4.7575	4.04	2.005	5.14	3.5225	**1.535**	1.9077E−10
Max Pareto front error	4.3375	3.1025	3.56	**2.215**	3.91	3.875	1.0060E−10
Generational distance	3.2475	1.955	5.06	**1.89**	3.6175	5.23	1.9471E−10
Inverted generational distance	3.0375	**1.86**	4.975	1.97	4.0725	5.085	2.2732E−10
Runtime	2.81	3.335	4.27	**1.19**	4.005	5.39	2.2589E−10

The best-performing algorithm in each metric is highlighted in bold

the solutions from converging to something good. Worth noting is that there is no apparent correlation between the size of the network and the hypervolume. This indicates that the network size alone is not a factor that allows us to predict the quality of the approximation sets.

Figure 10 shows the mean inverted generational distance achieved by each algorithm throughout its execution across all generated scenarios. It is interesting to note that while all of the algorithms are initialized with specifically tuned parameters they start and end at very similar values. Between 0 and 30% of function evaluations there is the largest disparity between algorithms, but after that they seem to converge towards a small set of desirable distance values.

Table 7 gives an overview of the statistical analysis carried out for each of the recorded metrics. The small p-values indicate that the results of the analysis are statistically significant for the group of algorithms under each metric. For each metric the following null hypothesis are not rejected:

Hypervolume: SPEA2 vs. AGE-II, NSGA-III vs. PESA-II
Max Pareto Front Error: AGE-I vs. AGE-II, SPEA2 vs. AGE-II, SPEA2 vs. AGE-I, NSGA-II vs. AGE-I, NSGA-III vs. SPEA2, NSGA-II vs. AGE-II
Generational Distance: NSGA-III vs. PESA-II, SPEA2 vs. AGE-II, NSGA-II vs. AGE-I
Inverted Generational Distance: NSGA-III vs. PESA-II, SPEA2 vs. AGE-II
Runtime: SPEA2 vs. AGE-I

It is important to note that while the metrics show that either NSGA-III or PESA-II is the best, except for max Pareto front error and runtime, none of the *post-hoc* procedures are able to reject the NSGA-III vs. PESA-II null hypothesis. Therefore, we are unable to definitively say whether one algorithm has performed better than the other. However, we are able to state that both NSGA-III and PESA-II outperformed the other algorithms.

6.3   Experiment 2: Inflicted Damage Analysis

In this experiment we examine the performance of the selected EMOO algorithms in a WSN comprised of 200 static nodes where an increasing number of them are labelled for replacement, e.g., in case of a malicious attack.

Concerning runtime, we can see a similar trend in Fig. 11 as the one depicted in Experiment 1. As the solution space increases in size we observe a smooth linear dependency between the percentage of nodes flagged for replacement and the time taken to compute a solution. SPEA2 and, to a much lesser extent, AGE-II suffers in terms of runtime when the solution space is small. This is likely due to the same reasons mentioned in our runtime discussion in Experiment 1.

Figure 12 shows that all of the algorithms perform similarly when a lower percentage of sensors needs to be replaced, but as that percentage increases the disparity of performance between algorithms appears. At damage levels above 20 % it can be seen that NSGA-III and PESA-II identify themselves as frontrunners for performance. This is likely due to the region-based selection method used by PESA-II and the method of selecting reference lines to drive intelligent exploration used by NSGA-III.

The results in terms of spacing are shown in Fig. 13. We can see that AGE-II and SPEA2 perform the best in terms of the variety of solutions that they present. The issue is that the quality of those solutions is not as high as those produced by other algorithms; this is especially true of AGE-II when comparing its results in Figs. 12 and 13. All of the other algorithms seem to have a similar amount of diversity among their resulting solutions which is interesting given their various selection methods.

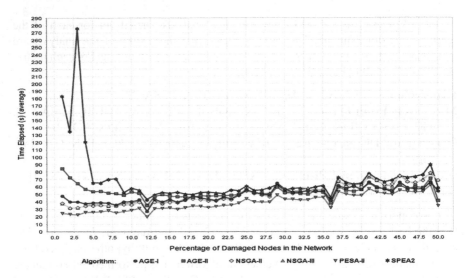

Fig. 11 Final execution results for time elapsed, in seconds, across all generated scenarios for the inflicted damage analysis experiment

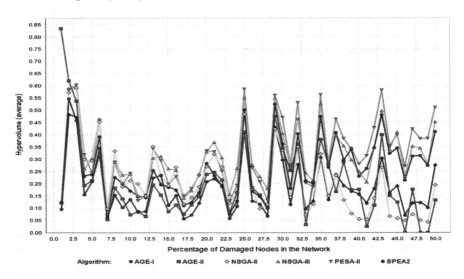

Fig. 12 Final execution results for hypervolume, across all generated scenarios for the inflicted damage analysis experiment

Figure 14 shows a snapshot of algorithm performance at various percentages of the execution budget, expressed as an average of the number of objective function evaluations across all scenarios. We can see that the majority of difference in performance comes before 30 % of function evaluations. It is interesting that each of

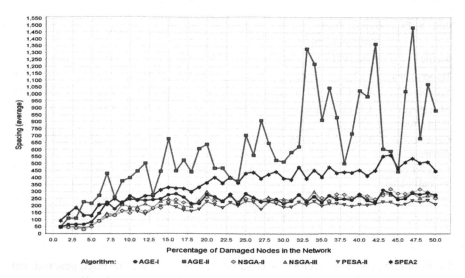

Fig. 13 Final execution results for spacing, across all generated scenarios for the inflicted damage analysis experiment

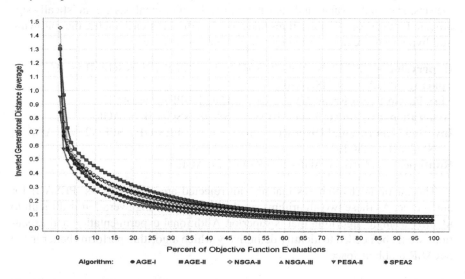

Fig. 14 Average results, across all generated scenarios, for the inverted generational distance for the inflicted damage analysis experiment

Table 8 Experiment 2: Friedman $N \times N$ test results

	Algorithm rank						
Metric	NSGA-II	NSGA-III	SPEA2	PESA-II	AGE-I	AGE-II	p-value
Hypervolume	3.8375	2.0425	4.55	**1.8825**	4.035	4.6525	1.6168E−10
Spacing	4.485	4.465	1.99	5.285	3.595	**1.18**	2.8214E−10
Max Pareto front error	4.1575	2.905	3.4775	**2.44**	3.16	4.86	9.6999E−11
Generational distance	3.075	**1.9125**	4.9	2.0475	3.24	5.825	2.6979E−10
Inverted generational distance	3.65	1.905	4.82	**1.885**	3.845	4.895	1.9804E−10
Runtime	3.075	3.875	5.825	**1.03**	3.315	3.88	2.0301E−10

The best-performing algorithm in each metric is highlighted in bold

the algorithms converge at close to the same narrow interval of inverted generational distance values. These results are near identical to those observed in Experiment 1.

Table 8 gives an overview of the statistical analysis found for each of the recorded metrics. The small p values indicate that the results of the analysis are statistically significant for the group of algorithms under each metric. For each metric the following null hypothesis are not rejected by any *post-hoc* procedure:

Hypervolume: SPEA2 vs. AGE-II, NSGA-III vs. PESA-II, NSGA-II vs. AGE-I
Spacing: NSGA-II vs. NSGA-III
Max Pareto Front Error: NSGA-III vs. AGE-I, SPEA2 vs. AGE-I
Generational Distance: NSGA-III vs. PESA-II, NSGA-II vs. AGE-I
Inverted Generational Distance: NSGA-III vs. PESA-II, SPEA2 vs. AGE-II, NSGA-II vs. AGE-I
Runtime: NSGA-III vs. AGE-II, NSGA-II vs. AGE-I

The important hypotheses that are not rejected are NSGA-III vs. PESA-II for any metric. As these two algorithms are the best performing overall, it is difficult to identify a single algorithm that is superior. Additional experimentation and executions would be needed to identify definitively which algorithm performs better in the RRASR problem space.

6.4 Experiment 3: Network Density Analysis

In this experiment we examine the effects of WSN topology on algorithm performance. To do this we generated solutions using the *MAX_DPA* method from [32]. We use distribution values ranging from 0 to 10 representing the minimum number of neighbors a sensor can have. We are interested in assessing the behaviour of the EMOO algorithms in both sparse and dense networks.

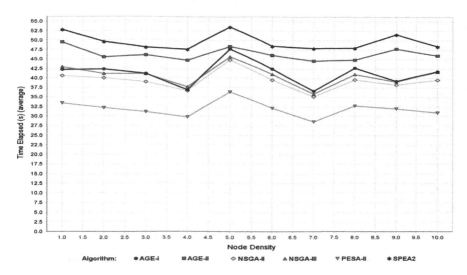

Fig. 15 Final execution results for time elapsed, in seconds, across all generated scenarios for the network density analysis experiment

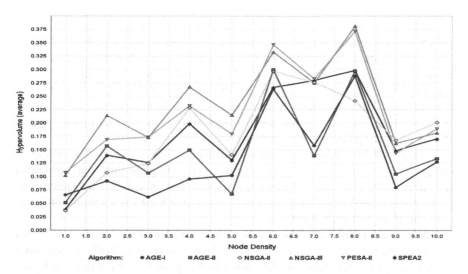

Fig. 16 Final execution results for hypervolume, across all generated scenarios for the network density analysis experiment

Figure 15 shows the runtime for each of the algorithms. This indicator remains similar across all network density values, therefore any disparity between algorithms is also highlighted. The same can be said for Fig. 16 where we examine the hypervolume of the EMOO algorithms at each network density value. We again see that the algorithms tend to perform similarly across each network density value.

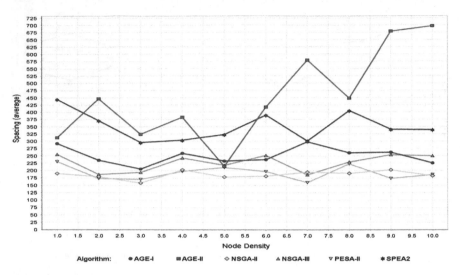

Fig. 17 Final execution results for spacing, across all generated scenarios for the network density analysis experiment

When considering spacing, Fig. 17 shows a different trend for AGE-II. While all the other algorithms tend to perform the same across node density values, AGE-II sets itself apart by having an upward trend as the node density value increases. This is an important find as it may indicate that the topology of the WSN has an effect on algorithm performance rather than just the size of the solution space, as demonstrated by Experiments 1 and 2. This is especially interesting as the EMOO algorithms have no way of noticing differences in network topology.

Looking at the mean inverted generational distance across all executions in Fig. 18 we see results much the same as Experiments 1 and 2. The algorithms show difference up to about 30 % of function evaluations and then trend towards the same point. This shows that each algorithm had managed to discover at least a few solutions that can be considered objectively good.

An overview of the statistical analysis for Experiment 3 is given in Table 9. We can see by the very small p-values that all of the results are statistically significant at the 5 % level. For each metric the following null hypothesis are not rejected by any *post-hoc* procedure:

Hypervolume: NSGA-III vs. PESA-II, AGE-I vs. AGE-II, NSGA-II vs. SPEA2, SPEA2 vs. AGE-II
Spacing: SPEA2 vs. AGE-II, NSGA-II vs. PESA-II, NSGA-III vs. AGE-I
Max Pareto Front Error: NSGA-III vs. AGE-II, NSGA-II vs. AGE-I, SPEA2 vs. PESA-II, NSGA-III vs. SPEA2, AGE-I vs. AGE-II, SPEA2 vs. AGE-II, NSGA-III vs. AGE-I, NSGA-II vs. AGE-II
Generational Distance: NSGA-III vs. PESA-II

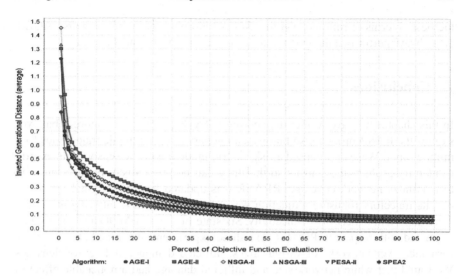

Fig. 18 Average results, across all generated scenarios, for the inverted generational distance for the network density analysis experiment

Table 9 Experiment 3: Friedman $N \times N$ test results

	Algorithm rank						
Metric	NSGA-II	NSGA-III	SPEA2	PESA-II	AGE-I	AGE-II	p-value
Hypervolume	3.4875	**1.775**	3.925	2	5.0875	4.725	8.1971E−11
Spacing	5.45	3.9	1.8	5.025	3.3	**1.525**	1.1685E−10
Max Pareto front error	4.7	3.375	2.7	**2.2**	4.375	3.65	5.0325E−10
Generational distance	2.75	**1.675**	4.925	1.75	3.9	6	9.3480E−10
Inverted generational distance	3.45	1.8	4.5	**1.625**	4.8	4.825	6.2807E−11
Runtime	2.1	3.25	6	**1**	3.675	4.975	1.0208E−10

The best-performing algorithm in each metric is highlighted in bold

Inverted Generational Distance: AGE-I vs. AGE-II, NSGA-III vs. PESA-II, SPEA2 vs. AGE-I, SPEA2 vs. AGE-II
Runtime: NSGA-III vs. AGE-I

We can notice some cases a null hypothesis with regards to the best performing algorithm is not rejected. This is the case for NSGA-III vs. PESA-II for hypervolume, generational distance, and inverted generational distance, for SPEA2 vs. AGE-II for spacing, and SPEA2 vs. PESA-II for max Pareto front error. In these cases we cannot definitively say that one algorithm has performed the best, only that some have performed better than others. We can also see that PESA-II always performed the best in terms of runtime, and that SPEA2 always performed the worst. AGE-II also always performed the worst in terms of generational distance.

7 Conclusions

In this chapter we have further defined the RRASR problem as a more robust version of the RASR problem, so that we can generate more reliable trajectories for a mobile robot. We also proposed a risk-centric modeling of the sensor units in order to identify problematic sensors so that they can be pro-actively replaced in order to maximize network coverage. RRASR was updated from its original version so that its trajectory robustness function was more accurate in identifying nodes that were important to network connectivity. Six state-of-the-art EMOO algorithms were evaluated with respect to RRASR. The three experiments examined algorithm performance over networks of various sizes, inflicted damage, and sensor densities. We found that while network size and inflicted damage had a noticeable effect on algorithm performance, network density did not. This is likely due to the stochastic, general nature of the selected algorithms as they do not make any changes based on network topography. Our results show that PESA-II and NSGA-III performed better than the other algorithms for all experiments. In the original work [4], it was found that PESA-II has the best performance. The presence of NSGA-III as a well performing algorithm is new, and likely due to the inclusion of parametric tuning, which has a better chance of giving all algorithms a suitable set of parameters as compared to a single set used for all algorithms. AGE-I and AGE-II were surprising in their poor performance given their success against the other algorithms using reference problems.

In the future we would like to expand problem to include solutions with multiple robots as well as relaxing hard constraints, such as the number of sensors that a robot can carry. We would also like to examine higher fidelity risk features such as battery risk which is represented by a simplistic model in our work but that could use more extensive models to describe such time-varying phenomena. In addition to higher fidelity risk features we would also like to examine higher fidelity data features such as transmission reliability modeled by the transmission rate as well as other factors (e.g. noise, interference, line of sight, environment) as well as physical reliability which can be modeled by age as well as other factors (e.g. Mean Time To Failure (MTTF), defect probability, harsh/extreme environments) and characteristics (e.g. accuracy, response time, precision).

References

1. Yick, J., Mukherjee, B., Ghosal, D.: Wireless sensor network survey. Comput. Netw. **52**(12), 2292–2330 (2008)
2. Falcon, R.: Towards Fault Reactiveness in Wireless Sensor Networks with Mobile Carrier Robots. PhD thesis, University of Ottawa, Ottawa, ON, Canada (2012)

3. Falcon, R., Li, X., Nayak, A., Stojmenovic, I.: The one-commodity traveling salesman problem with selective pickup and delivery: an Ant Colony approach. In: Proceedings of the IEEE Congress on Evolutionary Computation (CEC), Barcelona, Spain, pp. 4326–4333 (2010)
4. Desjardins, B., Falcon, R., Abielmona, R., Petriu, E.: A multi-objective optimization approach to reliable robot-assisted sensor relocation. In: 2015 IEEE Congress on Evolutionary Computation (CEC), pp. 956–964. IEEE (2015)
5. Falcon, R., Nayak, A., Abielmona, R.: An evolving risk management framework for wireless sensor networks. In: 2011 IEEE International Conference on Computational Intelligence for Measurement Systems and Applications (CIMSA) pp. 1–6. IEEE (2011)
6. Bringmann, K., Friedrich, T., Neumann, F., Wagner, M.: Approximation-guided evolutionary multi-objective optimization. In: IJCAI Proceedings-International Joint Conference on Artificial Intelligence, vol. 22, p. 1198 (2011)
7. Wagner, M., Neumann, F.: A fast approximation-guided evolutionary multi-objective algorithm. In: Proceedings of the 15th Annual Conference on Genetic and Evolutionary Computation, pp. 687–694. ACM (2013)
8. Lian-Ming, M., Xi-Li, D.: A novel ant colony system for solving the one-commodity traveling salesman problem with selective pickup and delivery. In: 2012 Eighth International Conference on Natural Computation (ICNC), pp. 1096–1101. IEEE (2012)
9. Falcon, R., Li, X., Nayak, A., Stojmenovic, I.: A Harmony-seeking firefly swarm to the periodic replacement of damaged sensors by a team of mobile robots. In: 2012 IEEE International Conference on Communications (ICC), (Ottawa, Canada), pp. 6436–6440 (2012)
10. Magklara, K., Zorbas, D., Razafindralambo, T.: Node discovery and replacement using mobile robot. In: Ad Hoc Networks, pp. 59–71. Springer, Berlin (2013)
11. Wang, Y., Barnawi, A., De Mello, R.F., Stojmenovic, I.: Localized ant colony of robots for redeployment in wireless sensor networks. Multi-Valued Logic Soft Comput. **23**, 35–51 (2014)
12. Fletcher, G., Li, X., Nayak, A., Stojmenovic, I.: Randomized robot-assisted relocation of sensors for coverage repair in wireless sensor networks. In: 2010 IEEE 72nd Vehicular Technology Conference Fall (VTC 2010-Fall), pp. 1–5. IEEE (2010)
13. Miao, Y., Yu-Ping, W.: Coverage repair strategies for wireless sensor networks using mobile actor based on evolutionary computing. Bull. Electr. Eng. Inform. **3**(3), 213–222 (2014)
14. Li, H., Barnawi, A., Stojmenovic, I., Wang, C.: Market-based sensor relocation by robot team in wireless sensor networks. Ad Hoc Sens. Wirel. Netw. **22**, 259–280 (2014)
15. Liao, X.-L., Ting, C.-K.: An evolutionary approach for the selective pickup and delivery problem. In: 2010 IEEE Congress on Evolutionary Computation (CEC), pp. 1–8. IEEE (2010)
16. Huang, Y.-H., Ting, C.-K.: Genetic algorithm with path relinking for the multi-vehicle selective pickup and delivery problem. In: 2011 IEEE Congress on Evolutionary Computation (CEC), pp. 1818–1825 (2011)
17. Liao, X.-L., Ting, C.-K.: Evolutionary algorithms using adaptive mutation for the selective pickup and delivery problem. In: 2012 IEEE Congress on Evolutionary Computation (CEC), pp. 1–8. IEEE (2012)
18. Bruck, B.P., dos Santos, A.G., Arroyo, J.E.C.: Hybrid metaheuristic for the single vehicle routing problem with deliveries and selective pickups. In: 2012 IEEE Congress on Evolutionary Computation (CEC), pp. 1–8. IEEE (2012)
19. Bruck, B.P., Santos, A., Arroyo, J.: An evolutionary algorithm and a variable neighborhood descent algorithm for the single vehicle problem with deliveries and selective pickups. In: Proceedings of the 2012 CLAIO/SBPO, Rio de Janeiro, Brazil (2012)
20. Ting, C.-K., Liao, X.-L.: The selective pickup and delivery problem: formulation and a memetic algorithm. Int. J. Prod. Econ. **141**(1), 199–211 (2013)
21. Bruck, B.P., dos Santos, A.: Hybrid approach for the multiple vehicle routing problem with deliveries and selective pickups. In: 2012 12th International Conference on Hybrid Intelligent Systems (HIS), pp. 265–270 (2012)
22. Falcon, R., Abielmona, R.: A response-aware risk management framework for search-and-rescue operations. In: 2012 IEEE Congress on Evolutionary Computation (CEC), pp. 1–8. IEEE (2012)

23. McCausland, J., Di Nardo, G., Falcon, R., Abielmona, R., Groza, V., Petriu, E.: A proactive risk-aware robotic sensor network for critical infrastructure protection. In: 2013 IEEE International Conference on Computational Intelligence and Virtual Environments for Measurement Systems and Applications (CIVEMSA), pp. 132–137. IEEE (2013)
24. McCausland, J., Abielmona, R., Falcon, R., Cretu, A.-M., Petriu, E.: Auction-based node selection of optimal and concurrent responses for a risk-aware robotic sensor network. In: 2013 IEEE International Symposium on Robotic and Sensors Environments (ROSE), pp. 136–141. IEEE (2013)
25. Falcon, R., Abielmona, R., Billings, S., Plachkov, A., Abbass, H.: Risk management with hard-soft data fusion in maritime domain awareness. In: 2014 Seventh IEEE Symposium on Computational Intelligence for Security and Defense Applications (CISDA), pp. 1–8. IEEE (2014)
26. Falcon, R., Abielmona, R., Billings, S.: Risk-driven intent assessment and response generation in maritime surveillance operations. In: 2015 IEEE International Inter-Disciplinary Conference on Cognitive Methods in Situation Awareness and Decision Support (CogSIMA), pp. 151–157. IEEE (2015)
27. Mamdani, E.H., Assilian, S.: An experiment in linguistic synthesis with a fuzzy logic controller. Int. J. Man–Mach. Stud. 7(1), 1–13 (1975)
28. Chen, J., Kher, S., Somani, A.: Distributed fault detection of wireless sensor networks. In: Proceedings of the 2006 Workshop on Dependability Issues in Wireless Ad Hoc Networks and Sensor Networks, pp. 65–72. ACM (2006)
29. Golumbic, M.C.: Algorithmic Graph Theory and Perfect Graphs, 2 edn, p. 2. Elsevier, Amsterdam (2004)
30. Deb, K., Pratap, A., Agarwal, S., Meyarivan, T.: A fast and elitist multiobjective genetic algorithm: NSGA-II. IEEE Trans. Evol. Comput. 6, 182–197 (2002)
31. Goldberg, D.E., Lingle, R.: Alleles, loci, and the traveling salesman problem. In: Proceedings of an International Conference on Genetic Algorithms and Their Applications, vol. 154. Lawrence Erlbaum, Hillsdale (1985)
32. Onat, F.A., Stojmenovic, I., Yanikomeroglu, H.: Generating random graphs for the simulation of wireless ad hoc, actuator, sensor, and internet networks. Pervasive Mob. Comput. 4(5), 597–615 (2008)
33. Deb, K., Jain, H.: An evolutionary many-objective optimization algorithm using reference-point-based nondominated sorting approach, part I: solving problems with box constraints. IEEE Trans. Evol. Comput. 18, 577–601 (2014)
34. Zitzler, E., Laumanns, M., Thiele, L., Zitzler, E., Zitzler, E., Thiele, L., Thiele, L.: SPEA2: Improving the Strength Pareto Evolutionary Algorithm. Eurogen, vol. 3242, pp. 95–100 (2001)
35. Corne, D.W., Jerram, N.R., Knowles, J.D., Oates, M.J., et al.: PESA-II: region-based selection in evolutionary multiobjective optimization. In: Proceedings of the Genetic and Evolutionary Computation Conference (GECCO2001, pp. 283–290. Morgan Kaufmann, Los Altos (2001)
36. Laumanns, M., Thiele, L., Deb, K., Zitzler, E.: Combining convergence and diversity in evolutionary multiobjective optimization. Evol. Comput. 10(3), 263–282 (2002)
37. Coello, C.C., Lamont, G.B., Van Veldhuizen, D.A.: Evolutionary Algorithms for Solving Multi-objective Problems. Springer, Berlin (2007)
38. Sierra, M.R., Coello, C.A.C.: Improving pso-based multi-objective optimization using crowding, mutation and-dominance. In: Evolutionary Multi-criterion Optimization, pp. 505–519. Springer, Berlin (2005)
39. Saltelli, A., Ratto, M., Andres, T., Campolongo, F., Cariboni, J., Gatelli, D., Saisana, M., Tarantola, S.: Global Sensitivity Analysis: The Primer. Wiley, New York (2008)
40. Derrac, J., García, S., Molina, D., Herrera, F.: A practical tutorial on the use of nonparametric statistical tests as a methodology for comparing evolutionary and swarm intelligence algorithms. Swarm Evol. Comput. 1(1), 3–18 (2011)

Printed in the United States
By Bookmasters